THE
COMPLEX MATTERS
OF THE MIND

STUDIES OF NONLINEAR PHENOMENA IN LIFE SCIENCE

Editor-in-Charge: Bruce J. West

Studies of Nonlinear Phenomena in Life Science – Vol. 6

THE
COMPLEX MATTERS
OF THE MIND

Editor

Franco Orsucci

Department of Psychology, Rome International University
and Institute for Complexity Studies, Rome

 **World Scientific**
Singapore • New Jersey • London • Hong Kong

Published by

World Scientific Publishing Co. Pte. Ltd.

P O Box 128, Farrer Road, Singapore 912805

USA office: Suite 1B, 1060 Main Street, River Edge, NJ 07661

UK office: 57 Shelton Street, Covent Garden, London WC2H 9HE

Library of Congress Cataloging-in-Publication Data
The complex matters of the mind / editor, Franco Orsucci.
     p. cm. -- (Studies of nonlinear phenomena in life science ; vol. 6)
    Based on contributions to the International Conference on "Chaos,
Fractals, and Models," held at the University of Pavia in Nov. 1996.
    Includes bibliographical references.
    ISBN 9810233396
    1. Mind and body -- Congresses.   2. Brain -- Psychophysiology --
Congresses.   3. Nonlinear systems -- Congresses.   4. Chaotic behavior
in systems -- Congresses. I. Orsucci, Franco. II. International Conference
on "Chaos, Fractals, and Models" (1996 : University of Pavia)   III. Series.
BF161.C636   1998
153--dc21   98-2516
                                                  CIP

British Library Cataloguing-in-Publication Data
A catalogue record for this book is available from the British Library.

Cover design and drawing by Sharon S. West.

Printed in Singapore by Uto-Print

CONTENTS

Nonlinear Dynamics in Language and Psychobiological Interactions
F. Orsucci

Minimal Models for Dyadic Processes: A Review
S. Rinaldi and A. Gragnani

**Fractal Dynamics of Heartbeat Interval Fluctuations in
Health and Disease**
*M. Meyer, C. Marconi, A. Rahmel, B. Grassi, G. Ferretti,
J. E. Skinner, N. Cerretelli*

Epistemological and Treatment Implications of Nonlinear Dynamics
A. H. Stein

**The Six Fundamental Characteristics of Chaos and their
Clinical Relevance to Psychiatry: A New Hypothesis for the
Origin of Psychosis**
G. B. Schmid

INTRODUCTION

"What kind of material process is directly
associated with consciousness?"
Erwin Schrödinger[1]

This book originated from the meetings of different scholars during the International Conference on "Chaos, Fractals and Models", which was held at the University of Pavia (Italy) in November 1996. This is the meditated result which follows our debate on the successes and difficulties of Nonlinear Studies, particularly in the areas of Mind Sciences. The editing of the book has been directed to offer a global reference in this field, which covers a wide range of possible approaches: theoretical, empirical, clinical, individual, social, biological and psychological. Every contribution starts from one of these areas, but tends to pass through the others. It is the effect of the necessary integration in the Mind Sciences, and this book should be considered as a step in this evolutionary process.

We had started with several questions: is an interdisciplinary contamination of complexity studies in different disciplines useful? Does this contamination originate a transdisciplinary toolbox of methods and models which is worth calling "Nonlinear Science"? What are the relations between the metaphoric approach and the mathematical approach in natural science and the humanities?

Complexity in the Life Sciences represents a fundamental workbench for these kinds of problems, and a fascinating challenge. One of the most promising fields, in these areas, is represented by studies on mind functioning.

Mind definition could be approached in the sense suggested by Bateson[2] as the name of a set of some functioning features of a living system. If an open system is an aggregate of different parts, which interact in cyclical loops, and modify itself by learning from experience; then this system can be called a *mind*. This means that every living system can function as a mind. From this vantage point, the human mind can be regarded beyond the Cartesian dualism *res cogitans* vs *res extensa*..

As suggested by Winnicott[3] "mind" becomes a peculiar mode in the functioning of the global *psyche-soma* being: mind *is* (in) the body. And mind is also a group of different subjects which collaborate to build a collective common sense.[4]

Nonlinear Dynamics represent both the essential framework and the basic toolbox for these studies.

Rome, September 1997   FRANCO ORSUCCI
                                          Rome International University

Acknowledgments

Francesco Corrao inspired some of the ideas for the origins of this book. Lydia Pallier was a source of reflections. Roberto Tagliacozzo originated clinical reconsiderations. Giorgio Parisi gave precious hints and trust. Graeme J. Taylor was our lead reference in psychosomatics. Kimberly Walter collaborated in formulating the style of the book. Kim Tan was an efficient and careful editorial contact at WSP.

References

1. E. Schroedinger, *Mind and Matter* (Cambridge, Cambridge University Press, 1958).
2. G. Bateson, *Mind and Nature* (New York, Dutton, 1979).
3. D.W. Winnicott, "Mind and its relations to the psyche-soma", *British Journal of Medical Psychology* **27**, 1954.
4. W.R. Bion, *Transformations: Change from Learning to Growth* (London, Heinemann, 1965).

FOREWORD

WALTER J. FREEMAN
University of California at Berkeley

The central concern of brain science since Hippocrates 2,500 years ago has been the relations between mind, body and brain. The ancient Greek ideal of the healthy mind in a healthy body stemmed in part from the recognition that the body was composed of the same four humors that comprised the rest of the sublunar world, and in part from the experience of harmony between the spiritual and material worlds that came with righteous living, as codified in their exuberant varieties of philosophy and natural science.

How could it happen that the flux and balance of the humors combining in the heart and brain yielded all of the impressions, memories, feelings of beauty, night terrors, and exaltations of the spirit in love, war, and games that constitute the lives of humans? The capacities for abstraction and self-awareness early on gave an enduring answer in the form of a separation between the material and spiritual worlds, which was brought to a pinnacle of elaboration in the philosophy of Plato. It was carried through the Middle Ages under the cloak of Christianity and was re-born in modern times through Descartes. Mind was the operation of the soul through the brain and body, like the pilot of a boat in the Cartesian machine metaphor. The muscles were worked by the flow of animal spirits from the ventricles through the nerves in analogy to the flow of blood through capillaries.

The most significant departure from this answer came in the middle of the 19th century, when the Young Turks under Helmholtz and Sechenov broke with their mentor, Johannes Müller, said to be the last of the great vitalists, by declaring that all physiological phenomena were to be explained only in terms of physics and chemistry without resort to spiritual forces. They did not deny the existence of the soul, but they refused to appeal to it as a source of explanation. Their centerpiece was the discovery of the action potential, the "negative variation", by Du Bois-Reymond, and the measurement of the conduction velocity of axons, which overthrew Müller's dictum that, being spiritual, the transmission was instantaneous. Their work culminated in the concept of nerve energy as the basis for all operations of the mind and brain, which were to be described in terms of Newtonian dynamics. The differential calculus that supported the burgeoning physical and engineering sciences was to be put into the service of neuropsychiatrists by the experimental biologists already probing brains with electrodes and chemical substances, seeking answers to the questions of how meaning, awareness, and consciousness might arise from energy states.

This bold hypothesis came to a bad end. The supposedly rigorous scientific studies of hysteria by Charcot became a laughing stock, when his dramatic findings were shown to manifest the power of suggestion in susceptible patients. J. Hughlings Jackson abandoned his attempts to define resistances in brain pathways. Freud turned his back on thermodynamics and went into literature. The behaviorists chose to regard mind as an epiphenomenon not worthy of scientific attention. Most significantly, nerve energy was shown not to conform to the 1st law of thermodynamics.[1] It is not conserved. Brains, with their infinite sink of waste heat and entropy in the venous outflow, are not constrained by the 2nd law. Whether brains are waking or sleeping makes little difference in their physical energy consumption.

By the middle of the 20th century a new hypothesis arose, whereby flows of information replaced the flows of energy. Neurons were conceived first as digital switches in networks capable of doing Boolean algebra, and later as microprocessors performing complex operations on the information brought to them on sensory pathways. With the help of increasingly powerful digital computers for constructing and solving the network equations, investigators have developed the new field of computational neuroscience. Neurons in brains are conceived to generate space-time patterns of action potentials, which serve as symbols to represent sensations, perceptions, retrieved memories, and commands for complex behaviors. Once again, the questions have come forth, of how meaning is attached to the firings of neurons, how qualia can be accounted for, and, above all, how does consciousness arise and operate? Titans are in battle, but no answers have been forthcoming.

Meanwhile, like primitive mammals among dinosaurs, a new breed of neuroscientists has been flourishing, who have discovered a fundamentally new and vital approach to the age-old question. This is a return to the field of dynamics as the source of tools, but with a highly significant twist. Whereas the Newtonian dynamics that has dominated physics and biology for several centuries is rigid, deterministic, and precisely predictable, the new field of nonlinear dynamics opens a vast field of complexity to exploration and modeling. The key concept is self-organization. Given an adequate supply of energy and a sink for waste disposal, a collection of interacting elements such as molecules, neurons, organs or people can create new structure from within. They need not depend on genes or the environment for instructions on how to behave. They proceed in accordance with internal logics that express hidden potentials for both expected and unexpected patterns having unbounded complexity. The differences between brains and extant devices can be characterized as follows.[1]

COMPARISON OF ARTIFICIAL NEURAL NETWORKS AND BRAINS

A.N.N.	BRAINS
Stationary	Unstable
Input-driven	Self-organizing
Task-oriented	Goal-oriented
Schematic	Hierarchical
Computational	Dynamical
bits	flows
symbols	patterns
information	meaning

MEMORY	REMEMBERING
An object	A process
Representation	Trajectory
Gradient descent	State transition
Retrieval	Construction

TEST BY:	
internal match	action into world

New concepts and a new set of tools are needed that in principle match the complexities of mind, brain and behavior. It is not the case that soul or spirit is to be explained away, but that it is not to be used to avoid confronting the problem of self-control and free will. Likewise, panpsychism[2-4] enervates the resolution needed to explore the dynamics of complex material systems that construct their own goals and pursue them with flexibility and resourcefulness. The adoption of belief in regard to supernal entities is a matter of individual conscience, whereas the new dynamics offers an opportunity to look at the relations from a fresh point of view. Best of all, entry into the field is open to everyone. Advanced training in mathematics and the physical sciences is not required. All that is needed is an open mind, a readiness to play with ideas in the spirit of the gifted amateur, and a willingness to fail on first attempts without becoming discouraged.

The essays in this book are examples of brave efforts to take up these new tools, explain their use, and apply them to significant problems by building models.

Arecchi opens the dialog with an essay on the epistemic processes of coming to terms with complexity and complex systems. Models are linguistic devices using a variety of languages, each with its set of rules for defining and using symbols, and its semiotic difficulties in assigning meaning with respect to the properties of that which is being modeled. He shows how sequences of metaphors, which are the content and medium of thinking (Lakoff 1987), are transformed from classical logic into distributions by use of fuzzy logic, and then into an evolving semantic space by use of time-dependent nonlinear dynamics.

Pietronero reports how many experiments and observations suggest that many physical systems develop spontaneously selforganizing scale invariant behaviors. Pattern formation, aggregation phenomena, biological and geological systems, disordered materials, clustering of matter in the universe are just some of the fields in which scale invariance in space and/or time has been observed as a common basic feature. The value of the impact of these concepts, however, is still rather controversial.

Sulis broadens the dialog by tackling that most human of attributes, the use of language. He describes the limitations of natural and formal languages particularly in relation to the analysis of behavior, and introduces nonlinear dynamics as a new language. He raises questions regarding what is its relationship to prior languages, and which dynamical properties can be used to model intrinsic linguistic structure. He proposes that answers may be found through the construction of dynamical automata, which would by their behavior provide functional evidence of an elementary understanding of the neural mechanisms by which thinking is done.

Orsucci extends the modeling more deeply into psychosexual behavior and "the biology of relations", combining insights from psychiatry with schemata of the architecture of human brains and the necessary incorporation of emotions by the autonomic and neuroendocrine systems, a refreshing reminder that brains exist only in and through bodies, and that the "brain in a vat" is chimera of interest only to the artificial intelligentsia.[5]

Rinaldi and Gragnani exemplify this utility of nonlinear models for restructuring well-known situations and interacting characters involved in dyadic love-hate relationships to engender surprising new points of view and to make interesting and provocative simplifications. Their equations consider the processes involving pleasure, appeal, yielding a positive linear system in which appeal plays the central role in community structure. They extend their model to include differing types of personality, and develop it further by describing the dynamics of Petrarch's love affair with a married woman, Laura. Petrarch appears to have avoided the chaos that would have ensued from the instability of the classical three-body problem by not mentioning the husband.

Meyer et al. introduce some of these powerful tools — dimensional analysis, multidimensional scaling, fractals, periodic and aperiodic oscillations, and the concepts of stationarity and autonomy — in the biological context of the analysis of information provided by the electrocardiographic measurement of the intervals between heartbeats, a most direct step from the physical sciences into biology and the behavioral sciences, a kind of testing ground for the management of the metaphorical transfer, in anticipation of the greater complexities that lie ahead. Most useful is their comparison of the rhythm of the denervated heart after transplantation, because the greater part of the normal variation reflects the participation of the brain. This approach can be applied to many other autonomic and musculoskeletal measurements.

Stein makes a frontal assault on the pretensions of psychoanalysis to be an exact science, that collapse in the face of the brutally nonlogical, unpredictable, chaotic behavior of human beings. Yet there is intelligent structure in seemingly irrational actions, which he characterizes inter alia as "play", and which he shows can be captured by descriptions based in nonlinear dynamics.

Schmid goes further in the modeling of pre-rational, pre-logical behavior as it appears in psychosis. He compares and contrasts classical, linearly causal, object-oriented psychiatry with the new nonlinear, circularly causal, process-oriented psychiatry, in terms of the characterization of dynamical disease as an abnormal form of chaos, leading to new opportunities for diagnosis, prediction, and especially control as a means for treatment of illness. In particular he proposes that, although the developmental course of psychosis may be chaotic, the mental state of psychosis may be a linearized information processing pathology. This is indeed a challenging prospect, at the forefront of the theory of chaos, where the possibilities are being explored by mathematicians and engineers to regulate the form and degree of chaos in complex systems, and adapt them for human purposes.[6-8]

Smith applies the theory of chaos to the self-organizing properties of social communities, noting its affinities to reflexive sociology and game theory, and offering as examples his studies of grassroots organizations in small Italian towns. This direction of intellectual inquiry is opportune and of great value, because paleoanthropological and comparative neurobiological evidence of the past few decades has made it clear that the human brain has evolved primarily as an organ of social integration.[1] No better evidence exists than the fact that the anterior half of the human cerebrum, the frontal lobes, is the locus of the mechanisms of insight and foresight for the genesis of social behavior. Whereas posterior lesions lead to blindness, deafness, and loss of body image, frontal lesions lead to social blindness, the loss of those delicate feelers that enable humans to understand each other and engage in cooperation to mutual satisfaction.

Giuliani concludes the set of chapters making explicit the hazards as well as the fecundity of using models from the physical sciences as tools to re-formulate and heuristically explore the domains of "soft sciences like economy, psychology and biology". Cold steel and metal oxides are not intrinsic to organic bodies and psychological relationships, yet the analyses of complex relationships by dissection with the scalpel and retention with the tape recorder are enormously facilitated by use of these tools. Giuliani provides a valuable service in recognizing that models are metaphors to be used at different levels in a hierarchical view of mind-brain-body, but not to be taken too literally, or too seriously.

Taken together, these chapters reveal the power of new concepts and techniques from nonlinear dynamics not to unify the sciences but to give the degree of abstraction that is necessary to think about disparate fields of intellectual inquiry, and in the same paragraph to write about body, brain, mind and society without embarrassment. Yet a note of caution is needed. Models are neither true nor false. They help to sharpen the terms of analyses, to open unexpected vistas and connections, and to pose new questions and hypotheses.[9] The worst mistake a modeler can make is to regard his or her model as a statement of truth, or as anything more than a heuristic probe, a scaffold, or a bridge that may or may not lead to an established theory. We desperately need the models, but we cannot rest with them. The materials in these chapters serve toward filling that need and stimulate further efforts to understand the multifaceted self-organizing properties of this remarkable entity: body-brain-mind-society.

References

1. W.J. Freeman, *Societies of Brains. A Study in the Neuroscience of Love and Hate.* Hillsdale NJ: Lawrence Erlbaum Associates, 1995.
2. A.N. Whitehead, *Modes of Thought.* New York: Macmillan 1938.
3. R. Penrose, *Shadows of the Mind.* Oxford UK: Oxford University Press, 1994.
4. D.J. Chalmers, *The Conscious Mind. In Search of a Fundamental Theory.* New York: Oxford University Press, 1996.
5. A. Clark, *Being There. Putting Brain, Body, and World Together Again.* Cambridge, MA: MIT Press,1997.
6. W.L. Ditto, S.N. Rauseo, M.L. Spano, Experimental control of chaos. *Physical Review Letters* **26**: 3211–3214, 1990.
7. L.M. Pecora, T.L. Carroll, Pseudoperiodic drifting: Eliminating multipole domains of attraction using chaos. *Physical Review Letters* **67**: 945-948, 1991.
8. W.J. Freeman, H-J. Chang, B.C. Burke, P.A. Rose, J. Badler, Taming Chaos: Stabilization of aperiodic attractors by noise. *IEEE Transactions: Special Issue on the Control of Chaos.* 1997, in press.
9. G. Lakoff, *Women, Fire, and Dangerous Things : What Categories Reveal About the Mind.* Chicago : University of Chicago Press, 1987.

COMPLEXITY IN SCIENCE: SYNTAXIS VERSUS SEMANTICS

F.T. ARECCHI

Istituto Nazionale di Ottica, Largo E. Fermi, 6
50125 Firenze
E-mail: arecchi@firefox.ino.it

A definition of the objects of a science in terms of precise measuring operations M gives the objects a set-theoretical character whereby complexity, seen as a multiplicity of possible final outcomes, emerges. An adaptive strategy introduces a frequent readjustment of the M settings, which reduces that multiplicity. This way, an adaptive cognitive task can be seen as the extraction of a simple map out of a complex landscape.

1. Introduction

First of all I will distinguish between A) "Complexity" and B) "Complex Systems".

The first term has been differently defined in formal languages [1], computer science [2] and nonlinear signal analysis [3], starting in the early 80's with the intrinsic non predictability associated with chaotic time series [4].

A) is associated with epistemic processes. Once a time series of data, coded in a given alphabet, has been assigned, can one retrieve the meaning of the message just by perusal of that sequence? Is "meaning" just i) discovering the grammatical rules which allow some symbol sequences (words) and forbid some other ones, or also ii) attributing a likelihood of occurrence to each word and hence attempting predictions about the future of the time series?

This A) approach has received different formulations, with different solutions leading to automatic procedures on complexity assignment [5-9].

On the other hand, natural scientists have rather focussed on the fact that reality uncovers lots of complex structures. It is common experience that the perception of an event can be expressed in many ways, not mutually reducible to each other, and this may be the token of a complex system.

Two different tasks are then faced by investigators of A) and B) problems, namely,

A)   given an input, coded as a word sequence of a language, what is the optimal use we can make of it? We call "certitude" the subjective confidence that we have done the best in grasping the inner rules of that sequence;

B)   facing a piece of world, can we express our knowledge of it in a suitable language, i.e. encode phenomena into symbol sequences from some alphabet (which later will be analyzed as in A))?

As we see, B) is "prior" to A). It appears as the problem that any living being has to solve, and it is usually faced by adaptive strategies, which of course we can later formalize as linguistic procedures, but which arise at a prelinguistic level and even determine the same choice of the most appropriate language [10]. This more fundamental problem is that of   "Truth", defined as   "adjustment of our expectations to the changing world".

The two approaches are altogether different. In case A) a learning machine can be foreseen which automatizes the quest for complexity [7]. On the other hand, case B) hints at the crucial role of a pre-linguistic stage where we still have to decide how to encode the stream of perceived phenomena into a linguistic sequence, which is then exposed to the inquiry of the complexity machine A).

We do not enter the philosophical problems involved by this distinction. We just limit to saying that doing physics is B), that is, making it possible to encode our perceptions into a suitable language, not just building theoretical models to uncover rules and make predictions with regard to given sequences as A).

Our main conclusion is that while there may be a complexity machine for A), it is in principle impossible to introduce a science mächine for B). Hence B) remains a human endeavour not reducible to automatic procedures.

The paper is organized as follows.

Sec. 2 reviews current definitions of complexity showing the virtues and intrinsic limitations of a contextual analysis of a data stream (by "contextual" we mean that we relye on correlations and symmetries already built in the symbol sequence). Sec. 3 is a dynamic approach to a complex situation. While in Sec. 2 no apriori rules are requested but a data stream is preliminary, in Sec. 3 we take for granted the general rules of dynamics and prospect a variety of possible data streams, that is, a variety of different physical situations. This variety appears as a natural implementation of the intuitive concept of complexity, as it is representative of what we call complex systems.

In Sec. 4 we present an adaptive strategy recently introduced to recognize [11] and control [12] a chaotic dynamics. While an adaptive procedure was already incorporated in the complexity machine of Sec. 2, in order to fit the theory to the input sequence, here, in a more radical way, we change the same structure of the measuring apparatus M in order to provide different sets of measurement sequences to be later analyzed.

We conclude with some epistemological insights, recalling that a knowledge program based on assigned input data is how to make the best use of our mental representations according to a subjectivistic gnoseology started by Descartes. On the contrary, a knowledge based on readjustment of our measuring procedures appears as the most natural attitude of living beings.

2. Complexity of symbolic sequences

In computer science, we may define as complexity of a word (symbol sequence) some indicator of the cost implied in generating that sequence. There is a "space" cost (length of the instruction stored in the computer memory) and a "time" cost (the CPU time for generating the final result out of some initial instruction).

A space complexity C_1 [2] is defined as the length in bit of the minimal instruction which generates the wanted sequence. This indicator is <u>maximum</u> for a random number, since there is no compressed algorithm (that is, shorter than the number itself) to construct a random number.

A time complexity C_2, called "logical depth" (Bennett in Ref.5a) is defined as the CPU time required to generate the sequence out of the <u>minimal</u> instruction. C_2 is minimal for a random number, indeed, once the instruction has stored all the digits, just command: PRINT IT

Of course, for simple dynamical systems as a pendulum or the Newtonian two-body problem, both complexities are minimal.

While C_1 refers to the process of building a single item, C_2 corresponds to finding the properties of all possible outputs from a known source. Following Simon [10], C_1 refers to a process description and C_2 to a state description. Indeed C_1 corresponds to the effort to arrive at a specific object, and C_2 corresponds to the mental representation of the whole class of objects.

In fact, the exact specification of the final outcome is too much for the ambition of the natural scientist, whose goal is more modest. It may be condensed in the two following items:

i) to transmit some information, coded in a symbol sequence, to a receiver in a compact way, possibly economizing with respect to the actual string length, that is, making good use of the redundancies (this requires a preliminary study of the language style);

ii) to predict a given span of future, that is, to assign with some likelihood a group of forthcoming symbols;

For this second goal, introduction of a probability measure is crucial [6] in order to design a complexity-machine, able to make the best informational use of a given data set.

In view of the difference between A) and B) let us sketch the essential elements of knowledge building in natural sciences. First of all, we realize a device, or measuring apparatus M, whose output is informationally equivalent to what is going on in the observed piece of world.

We thus attribute to knowledge two different meanings:

i) as we face a phenomenon, M captures (presumably) the relevant aspects of it, so that we can transfer sufficient information, either to another partner or to ourselves if we have to reflect in order to build a possible theory. Knowledge improvement implies trying with different M apparatuses by a suitable program that we will explore in Sec. 4.

ii) As observer O_1 is exposed to a symbol stream, it has to transmit a compact explanation to O_2, so that O_2 is able to retrieve the same input sequence. The explanation consists of a tentative theory that we call model m.

The measuring apparatus M is characterized by the following elements[13]:

D:number of probes (dimensionality of the measurement space),

ε: resolution in the projected state space (total number of cells is ε^{-D}),

τ: time resolution,

$\beta = 1/T$ (T=noise temperature): fuzziness associated with a non sharp boundary of each resolution box, yielding some ambiguity in the assignment of an event to a specific cell of state space.

At any time slot of width τ, we extract ε^{-D} different space data, that we can encode in a suitable one dimensional string of symbols of an alphabet (e.g. binary). The modeler O_1 is inputted by some sequence s, and it sends an explanation which should enable O_2 to reconstruct an output s' = s.

Notice that M = M (D, ε, τ, β) is a whole class of possible instruments, and different individuals will give rise to different data sequences (different words) for the same input.

The explanation consists of a theoretical guess (model m) whose validity is tested by simulating an output and comparing it with the actual input data s. The difference yields an error signal e. Observer O_2 is provided with both m and e and it can reconstruct s' = s upon this information.

The virtue of the explanation X is to have a bit length $\|x\| = \|m\| + \|e\|$ much shorter than the sequence length $\|s\|$: this amounts to extract a relevant semantics out of the redundant features of s.

The explanatory machine is complex in so far as it spans over a whole class of models m. If one had access to a complete probabilistic description of the modeling universe, then the goal would be to maximize $\Pr(m \mid s)$, the probability of m conditional to the input s.

This ideal complete description is not available, but an approximation can be obtained by the Bayes' rule

$$Pr(m \mid s) = \frac{Pr(s \mid m)P(m)}{\sum_m P(s \mid m)}$$

Here $Pr(s \mid m)$ is the probability that a model m produces the given data and it is weighted by the probability $P(m)$ of the model class. Finally the normalization in the denominator depends only on the given data, and so can be dropped as a constant.

The most likely explanation corresponds to the shortest code of length $\|x\| = -\log_2 Pr(m \mid s)$.

We fix two criteria for a good explanation, namely:
i) x must explain s, that is, O_2 should resynthesize the original data s' = s;
ii) the bit length

$$\|x\| = \|m\| + \|e\|$$

must be minimized.

The efficiency of an explanation is given by the compression ratio

$$C(m,s) = \frac{\|x\|}{\|s\|} = \frac{\|m\| + \|e\|}{\|s\|}$$

C is a cost function. The optimal model minimizes this cost.

There are two limit cases. When the model is trivial ($\|m\| = 0$), the entire data are on the error channel: $\|e\| = \|s\|$. On the contrary a tautological model m=s has no error: $\|e\| = 0$.

The reconstruction of a state space out of an assigned time series and the assignment of transition probabilities among states is a task faced in many ways (see e.g. [5], [6]). According to the title of this section, here we have explored the complexity of a given symbolic sequence either from the computational point of view, aimed at reconstructing the single item, or from the probabilistic point of view, aimed at selecting a model within a class.

However, preliminary to that, the problem arises of how we have obtained a given sequence, and this implies a critical analysis of the measurement apparatus.

4. A dynamical approach to complexity

A theory is successful if it is a "compressed" description of the world, that is, if the length of its initial assumption is much shorter than a detailed description of the events themselves. At the start, a physical theory is just mathematics. It becomes a model, that is, it acquires semantic values, whenever we interpret the objects of the theory as elements of reality [14].

Therefore a scientific theory must be considered as a set of primitive concepts (defined by suitable measuring apparatuses M) related by axioms. The deduction of all possible consequences (theorems) provides predictions which have to be compared with the observations. If the observations falsify the expectations, then one tries with different axioms.

The deductive process is affected by a Gödel undecidability like any formal theory, in the sense that it is possible to build a well formed statement, but the rules of deduction are unable to decide whether that statement is true or false.

Besides that, a second drawback is represented by intractability, that is, by the exponential increase of possible outcomes among which we have to select the final state of a dynamical evolution. Fig. 1 sketches a bifurcation tree well familiar to computer scientists, as one has to perform a complex calculation with branch points implying multiple choices of the type "if-then". I rather consider it as the bifurcation tree of a complex nonlinear dynamics, as one changes a suitable control parameter α.

Going back to the reductionistic tentative of explaining reality out of its constituents, then we find an exponentially high number of possible outcomes, when only one is in fact that observed. This means that, while the theory, that is the syntaxis, would give equal probability to all branches of the tree, in reality we observe an organization process, whereby only one final state has a high probability of occurrence.

It is here necessary to recall some descriptive elements on the bifurcation of the stable branches of a nonlinear dynamics for different settings of a control parameter. These are the necessary ingredients of any complex dynamics. Notice that dynamical bifurcations in a system of interactive identical particles display specific symmetries (fig. 2a). Only external gradients break this symmetry (fig. 2b).

Thus, during the course of a dynamical evolution, either because some control parameters $\{\alpha\}$ are tuned from the outside to assume different values, or because internal feedbacks change the $\{\alpha\}$ set in course of time, starting from some initial conditions we expect an exponential increase of final states.

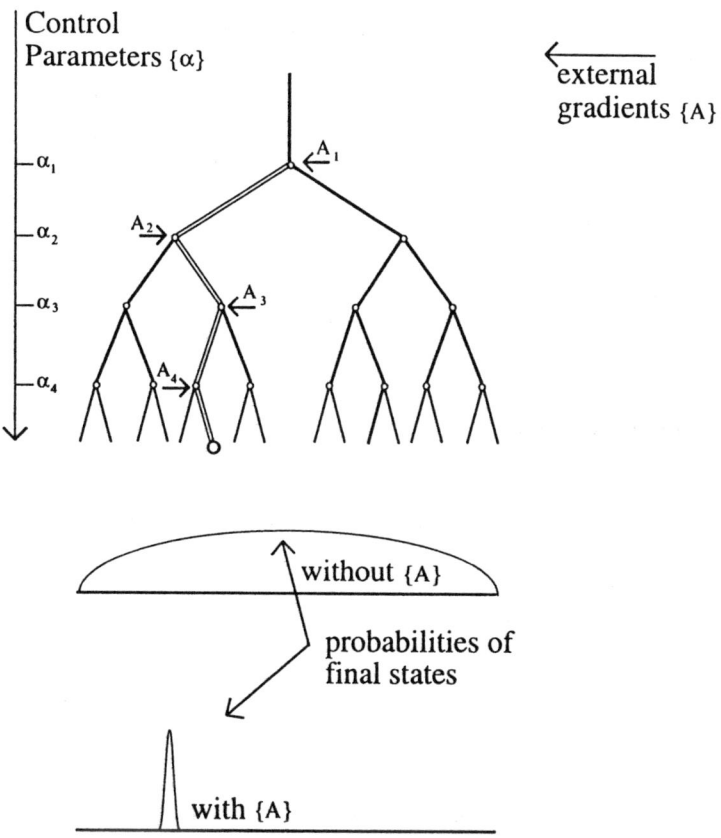

Control
Parameters {α}

external
gradients {A}

α_1

α_2

α_3

α_4

A_1

A_2

A_3

A_4

without {A}

probabilities of
final states

with {A}

Fig.1. Bifurcation tree in nonlinear dynamics. As a control parameter α is tuned
through different values, novel steady states appear. By tuning α from α_0 to
α_N, the system goes from 1 to 2^N different states. In the absence of external
gradients, all final outcomes are equally probable. We call "dynamical
complexity" such an ambiguity, and "organization" the occurrence of just one
event out of 2^N. This implies that at each α_i an external agent A_i has broken
the bifurcation symmetry (see next figure).

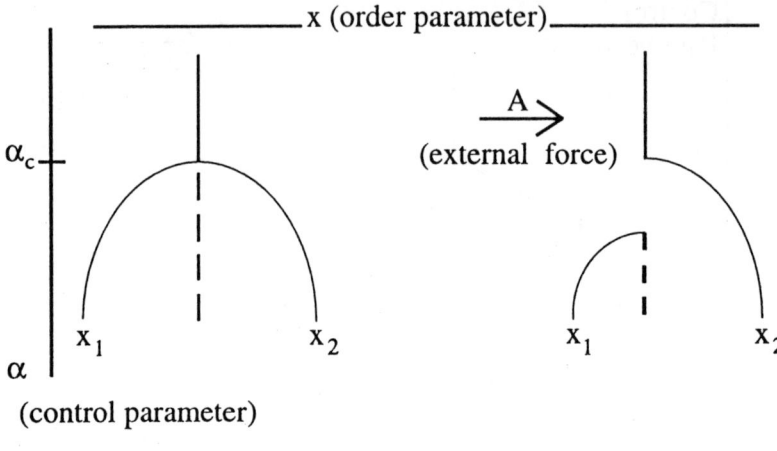

x (order parameter)

α_c

A (external force)

α (control parameter)

x_1 x_2 x_1 x_2

Probabilities $\alpha > \alpha_c$

$$P(x_1) = P(x_2) = 0.5 \qquad P(x_1|A) = 0 \qquad P(x_2|A) = 1$$

Fig.2. Examples of bifurcation diagrams. The dynamical variable x (order parameter) varies horizontally, the control parameter α varies vertically. Solid (dashed) lines represent stable (unstable) steady states as the control parameter is changed.
Left: symmetric bifurcation with equal probabilities for the two stable branches.
Right: asymmetric bifurcation in presence of an external field A. If the gap introduced by A between right and left branch is wider than the range of thermal fluctuations at the transition point α_c, then the right (left) branch has probability 1 (0).

Whenever there has been <u>organization,</u> this means that at each bifurcation vertex of fig. 1 the symmetry was broken by an agent external to the system under investigation.

We can thus stipulate the following things:

i) A set of control parameters

$$\alpha_1, \alpha_2, \ldots \alpha_N = \{\alpha\}$$

is responsible for successive bifurcations leading to an exponentially high number (in the example, of the order of 2^N) of final outcomes. If the system has no boundary effects (considered of infinite size) then all outcomes have comparable probabilities and we call complexity the impossibility of predicting which one is the state we will observe at the end of the chain of bifurcations.

ii) A set of external forces

$$A_1, A_2, \ldots A_N = \{A\}$$

applied at each bifurcation point break the symmetries, biasing toward a specific choice and eventually leading to a unique final state.

We are in presence of a conflict between (i) syntaxis represented by the set of rules (axioms) $\{\alpha\}$ and (ii) semantics represented by the intervening external agents $\{A\}$. The syntaxis provides 2^N legal outcomes. But if the system is open to the external world, the presence of which is expressed by $\{A\}$, then it organizes to a unique final outcome. Once the syntaxis $\{\alpha\}$ is known, the final result is therefore an ascertainment that the set of external events $\{A\}$ must have occurred. Therefore we can take $\{A\}$ as the element of reality in which our system is embedded.

We define "certitude" the correct application of the rules $\{\alpha\}$, and "truth" the adaptation to the reality which is expressed by $\{A\}$.

However the same final outcome would be reached by a different set of rules $\{\beta\}$. In such a case, retracing back the new tree of bifurcations, we would reconstruct a set $\{B\}$ of external agents. Thus, it seems that truth, $\{A\}$ or $\{B\}$, is language dependent!

Furthermore, the "emergence" of organization means that we can even build a set of axioms $\{\varepsilon\}$ which succeeds in predicting the correct final state without external perturbations, that is, $\{E\}=\varnothing$ (fig. 3). This is indeed the pretension of the so called "autopoiesis", or "self-organization" [15], to which I have opposed the term "hetero-organization" [16].

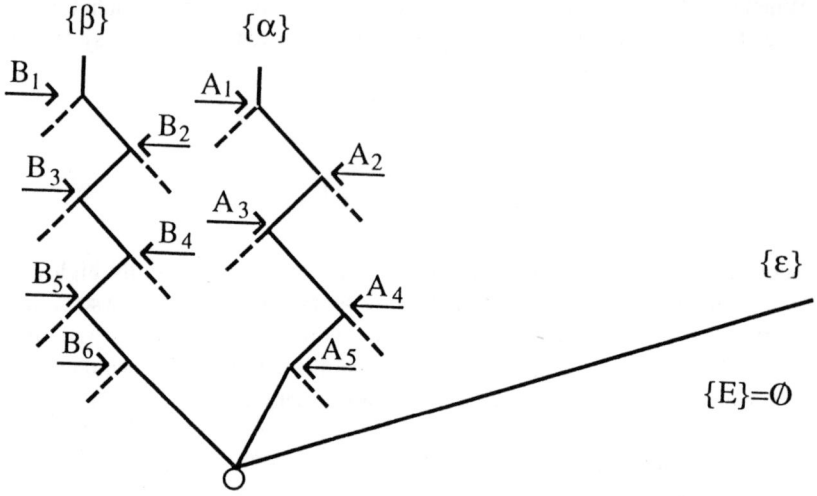

Fig.3. Different theoretical models may explain the same final state. The backward reconstruction of the dynamical history will then retrieve different classes of external agents.

From a cognitive point of view, the theory $\{\varepsilon\}$ can be reputed to be a "petitio principi", a tricky formulation tailored for a specific purpose and not applicable to slightly different situations. Rather than explicitly listing the elements of reality, as e.g. $\{A\}$ for $\{\alpha\}$, the user of language $\{\varepsilon\}$ has already exploited at a pre-formalized level the elements of reality, and has made good use of them in planning the axioms$\{\varepsilon\}$.

These pieces of knowledge which precede axiomatization have received different names, as "abduction" [17] or "tacit dimension" [18]. Some of them have been memorized as universal tools either in our genetic heritage, or during infancy in our learning age. This seems to be the cognitive valence of Jung's "archetypes" [19].

4. An adaptive measurement procedure

In Sec. 2 we introduced a measuring apparatus M (D, ε, τ, β) which can be modified by changing the number of probes D, space (ε) or time (τ) resolution, or fuzziness β. Any change in M leads to a different set of output numbers, and hence to different sequences s (words). We can specifically tailor the M characteristics in order to emphasize a given set of dynamical properties and project out some other ones. Since M acts as a projector into the subspace of the variables measured by M itself, changing M means changing the "point of view" under which we observe the world, and hence making a different theoretical model.

In the previous Section we show an indeterminacy in the reconstruction of the elements of reality {A, B, C, ...} which modify the dynamical theories {α, β, γ....} of a mentally isolated system. As a result, the truth is represented by a combination of a model developed for the subsystem, plus the external gradients, or boundary conditions, once the subsystem is embedded in a suitable environment.
Which one is the most appropriate among the pairs

$$\{\alpha, A\}, \{\beta, B\}, \{\gamma, C\}, \text{etc.?}$$

The question is equivalent to asking: among all possible measuring apparatuses M applicable to an event, which one is the most appropriate?
If we prefer not to decide, we independently make use of different M's and correspondingly define different sciences.

In real life, we have to face problems overlapping different separate sciences. For instance, a cardiac disease may be due to a global offset of the pace-maker, or to some local cytopathology or even to a drug effect acting at a biomolecular level.
Fig. 4 shows the difference between fixed and adaptive M. In the first case we sharply define three separate sciences. What can be exchanged among different specialists are not technical words, which are specific of each science, but just the residual metaphorical part, not filtered into the technical word. Should we say that two scientists of different areas always communicate by metaphors? A tentative way out (fuzzy logic [20]) is to avoid sharp definitions, so that different disciplinary terms have regions of overlap.

However the most natural approach seems to start with M at very low resolution, covering all the disciplinary areas, and then - depending upon a specific problem - to zoom toward one or the other narrow point of view.

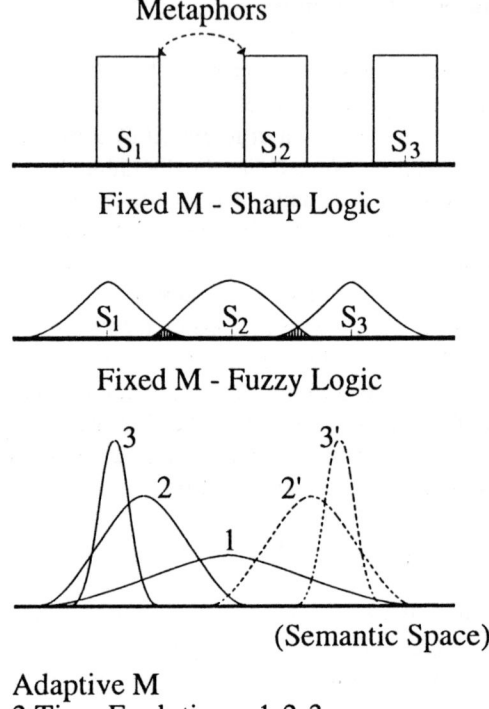

Metaphors

S_1 S_2 S_3

Fixed M - Sharp Logic

S_1 S_2 S_3

Fixed M - Fuzzy Logic

3   2   2'   3'

1

(Semantic Space)

Adaptive M
2 Time Evolutions: 1-2-3
                       1-2'-3'

Fig.4. Fixed measuring apparatuses giving rise to different sciences. Inter communication is based on metaphors (upmost figure) unless by fuzzy logic one accepts ill defined terms with overlaps (center figure).

When the measurement is no longer locked to a fixed position of semantic space, nor to a fixed resolution, then the scientific knowledge can cover, with different degrees of detail, different areas of the semantic space, allowing information exchange within a unique language.

A successful line of adaptive measurement has been worked out in my group[11-12]. It consists in developing a measuring procedure M whereby we observe a system only at "almost equal" geometric separations in state space. As a consequence, M is activated only for short time intervals at irregularly distributed times, separated from each other by unequal intervals. The stroboscopic sequence of these intervals has an information content which provides fast reliable answers to the following questions:

1. recognizing a chaotic dynamics (Lyapunov exponents, different unstable periodic orbits (UPO's);
2. discriminating determinism from stochastic noise;
3. controlling chaos, i.e. stabilizing one of the UPO's contained within a chaotic attractor.

When applied to the control of chaos[12], this adaptive algorithm is effective for values of the stroboscopic times much larger than the Runge-Kutta integration steps and smaller than the periods of all UPO's. In other words, the method introduces a natural adaptation time scale which is intermediate between the minimum resolution time of the dynamics and the time scale of the periodic orbits.

5. Conclusion

In this conclusion we discuss the truth value of scientific statements on the basis of the considerations of the previous Sections.

In the scientific investigation, we select a quantitative feature via an apparatus M. The emerging description of reality represents an observation "from one point of view" [21].

Due however to the variety of possible M's, we must justify at a metascientific level why we have selected that M rather than another one. This is a general question dealing with the role of those elements of reality which are preliminary to one particular program.

In Sec. 4 we have seen that adaptation is a slalom among different sets of rules $\{\alpha\}$, $\{\beta\}$,.... under the guidance of a preferential set of external elements which has non zero intersections with $\{A\}$, $\{B\}$... but does not coincide with any one of them.

In front of the truth problem we can take three attitudes, namely,

i) assume the adaptive strategy and its associated reality set as a kind of privileged reference frame. Indeed, being the result of an optimization process, it appears more appropriate than any particular theory, $\{\alpha\}$, or $\{\beta\}$etc.

ii) Consider the truth problem as a metatheoretical problem. At this metalevel, the set of all sets of truth values {A}, {B},..... has to be considered as the truth, but with the stipulation that any individual set makes sense only if associated with the corresponding theory.

iii) A more fundamental approach recovers the polysemicity of the ordinary language as a virtue, not a drawback. More than questioning the power of any specific theory we put into question the same set-theoretical approach to physics. We have seen that a fixed M provides a sharp connotation which allows to classify any observed entity within an appropriate set. Whence the set-theoretical character of all modern sciences, with the consequent antinomies of modern logic, transferred into the heart of the scientific language. An adaptive M means that the localization in semantic space is no longer as sharp as whenever it is defined by a precise stipulation as for the sets. This degree of smoothness lowers the accuracy but it releases the drawback of the antinomies.

Going back to the title, the truth versus certitude issue can be summarized by the following scheme (fig. 5). R stays for reality (whatever this means), S for a symbol interpreter (an intelligent being or even a Turing machine!), M is the measuring apparatus.

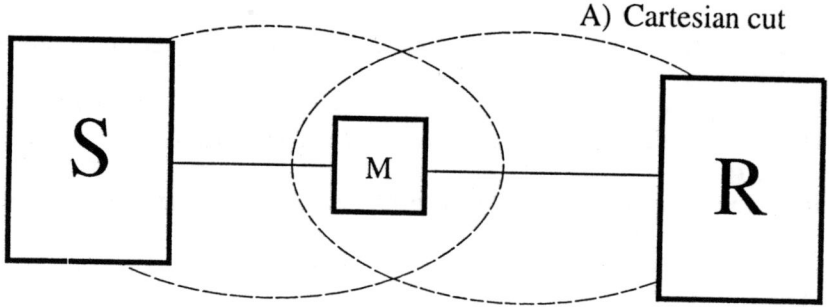

A) Cartesian cut

B) Realistic approach

Fig.5. Knowledge interpretation: R=reality, M=symbol generator (measuring apparatus), S=symbol interpreter (model builder).
Dashed line A (Cartesian cut): M+R provide representations as symbol sequences, which are interpreted by S. S can be replaced by Turing machine.
Dashed line B (realistic approach): S+M globally face R. Before producing outputs, M is readjusted among a class of possible measuring apparatuses by a prelinguistic procedure not expressible within the formal language which later M provides to S.

Up to now, M was not questioned, and the elaboration took place on the output of M. This was called as the <u>A) Complexity</u> approach, leading eventually to certitude, when the explanatory machine replicates exactly the input sequence with the minimal amount of information. From a gnoseological point of view, if M are our senses, as suggested by Hume, then S (called by Descartes "res cogitans") has to face <u>not the world, but the representation</u> already coded by M, that is, a grammatical problem inplementable by a machine. This means that Descartes' mind is equivalent to a Turing machine, as already suspected by many experts of Artificial Intelligence. This strong association of M on the side of R is equivalent to what Atmanspacher called "The Cartesian cut" [22].

On the contrary, the <u>B) Complex system</u> approach regards science as dealing with the world through an adaptive procedure M, for which however a linguistic foundation does not exist, because any linguistic formulation is subsequent to the operation of M. During the scientific operation, S+M act in an <u>entangled</u> way. A meta-level of investigation (psychology of cognition) is required to disentangle S from M.

In summary, there is a nonlinguistic residue in the scientific operation which then precludes a Turing machine from acting as a creative scientist.

16

References

1. J.E.Hopcroft and J.D.Ullman *"Introduction to automata theory, languages and computation"*, Addison-Wesley, Reading, MA (1979)

2a. A.N.Kolmogorov *"Three approaches to the quantitative definition of information"*, Problems of Information Trasmission **1**, 4, (1965).

2b. G.J.Chaitin *"On the length of programs for computing binary sequences"* J. Assoc. Comp. Math. **13**, 547 (1966).

3a. P.Grassberger, T.Schreiber and C.Schaffrath *"Nonlinear time sequences analysis"* Int. J. Bif. Chaos **1**, 521 (1991).

3b. H.D.I.Abarbanel, R.Brown,J.J.Sidorowich and L.S.Tsimring *"The analysis of observed chaotic data in physical systems"* Rev. Mod. Phys. **65**, 1331 (1993).

4. R.Shaw *"Strange attractors, chaotic behavior, and information flow"*, Z.Naturforsch., **36a**, 80 (1981).

5. See the following contributions in the series of Santa Fe Institute Proceedings Volumes:
   a)Ed. David Pines *"Emerging syntheses in science"* (1987) Vol.1
   b)Ed. W.H.Zurek *"Complexity, entropy and the physics of information"* (1990) Vol.8
   c)Ed.M.Casdagli and S.Eubank *"Nonlinear modeling and forecasting"* (1992) Vol.12

6. P.Grassberger *"Toward a quantitative theory of self-generated complexity"*, Int. J. Theor. Phys., **25**, 907 (1986).

7. J.P.Crutchfield and K.Young *"Inferring statistical complexity"* Phys Rev. Lett. **63**, 105 (1989).

8. G.D'Alessandro and A.Politi *"Hierarchical approach to complexity with applications to dynamic systems"* Phys Rev. Lett. **64**, 1609 (1990)

9. For an updated critical review see R.Badii and A.Politi *"Complexity: hierarchical structures and scaling in physics"*, Cambridge University Press, 1997, especially chapter 8 and 9.

10. H.Simon *"The architecture of complexity"*, Proc. Amer. Philos. Soc., **106**, 467 (1982).

11. F.T.Arecchi, G.F.Basti, S.Boccaletti and A.Perrone *"Adaptive recognition of a chaotic dynamics"* Europhys. Lett., **26**, 327 (1994).

12. S.Boccaletti and F.T.Arecchi *"Adaptive control of chaos"* Europhys. Lett. **31**, 127 (1995).

13. J.P.Crutchfield *"Semantics and thermodynamics"* in Ref.5c

14.   A. Tarski, *"The concept of truth in formalized languages"* in: Logic, semantics, mathematics: papers 1923-1938 by ATarski, Clarendon Press, Oxford,1956.

15.   W. Krohn, G. Küppers and H. Nowotny, *"Selforganization, Portrait of a scientific revolution"*, Kluwer Academic Publishers, (1990).

16.   F.T. Arecchi, *"A critical approach to complexity and self-organization"*, La Nuova Critica I-II, Quaderno 19-20, pp. 7-39, (1992).

17.   C.S. Peirce, *"Collected papers"*, Vol.s I-VI, Ed. by C. Harshorne and P. Weiss, Harvard University Press, Cambridge, Mass. 1931-35, Vol.s VII and VIII, Ed. by A.W. Burks, same publisher, 1958; see also: W. R. Hanson, *"Patterns of discovery: an inquiry into the conceptual foundations of science"*, Cambridge University Press, Cambridge, 1958.

18.   M. Polanyii, *"Personal knowledge: towards a post-critical philosophy"*, Routledge & Kegan Paul, London 1958.

19.   M.L. Von Franz, *"Psyche und Materie"*, Einsiedeln (C.H.) 1988.

20.   L. A. Zadeh, *"Fuzzy sets and applications"* (selected papers), J. Wiley, New York, 1987.

21.   E. Agazzi, *"Temi e problemi di filosofia della fisica"*, Ed. Abete, Roma, 1974 (Sec. 50)

22.   H. Atmanspacher, *"Complexity and meaning as a bridge across the Cartesian cut"*, J. of Consciousness Studies, 1, 168 (1994).

COMPLEXITY AND FRACTALS IN PHYSICS

LUCIANO PIETRONERO

ICTP, P.O. Box 586, 34100 – Trieste, Italy and
Dipartimento di Fisica, Università di Roma "La Sapienza",
Piazzale A. Moro 2, 00185 – Roma and INFM Unit of Roma1, Italy

Abstract

The study of complex systems and Fractal Geometry is having a profound impact on the physical sciences. These concepts change the way we look at nature and permit to include in the scientific domain new phenomena characterized by strong irregularities and self-similarity. Experiments and observations indeed suggest that many physical systems develop spontaneously self-similar behavior both in space and time. Pattern formation, aggregation phenomena, biological and geological systems, disordered materials, clustering of matter in the universe are just some of the fields in which scale invariance has been observed as a common basic feature. The value and impact of these concepts, however, is still rather controversial. In this paper we discuss the main advancements as well as the present limitations of these ideas by discussing some (representative) examples. We also outline the main open problems and the possible developments of the field.

1. Introduction: The Simple and the Complex

The traditional scientific approach is to consider the simplest element or question and study it in great detail. This approach focuses on the elementary "bricks" that constitute matter. From many of these studies it has been possible to derive some general laws of physics that apply in a much broader sense. So, the basic idea is that the knowledge of the properties of the individual constituents of matter allows us to understand also a large assembly of these elements. This is the reductionistic approach which has been very successful in many cases.

However, it is easy to imagine that for a cell or a living organism the situation is actually quite different. Cells and living organisms are made of atoms and we know a lot about the properties of individual atoms. But even if our knowledge of the atoms would be perfect it is clear that we could not understand the functioning of a cell. We are faced with a new situation: the knowledge of the properties of the individual elements are not sufficient to describe the entire structure. What happens is that these elements interact with each other, form complex structures and develop functions which have little to do with their isolated properties. In this respect we may represent this situation as the study of the "architecture" of matter and nature. It depends on the properties of the "bricks", but then it has its own characteristics and fundamental laws which cannot be related to those of the individual elements.

The simplest form of complex structures are those arising from the basic laws of physics but involve many interacting elements with irreversible dynamics and nonlinear interactions. Often, however, these structures arise from apparently simple rules. Examples are the familiar fractal structures like trees, coastlines, clouds, lightnings, turbulence in fluids and even the distribution of galaxies in the universe.[1] Also the case of deterministic chaos is an example of this type in which the iteration of a simple but nonlinear dynamical system leads to structures and behaviors of great complexity. In these cases complexity is often represented by the property of scale invariance of these strutures and this will be the main subject of this lecture. What is remarkable is that these properties arise spontaneously from the dynamics of the system without specific adjustments; in this sense they are self-organized.

A more complex but related situation is represented by the biochemistry of living systems. The laws of biology are related and based on those of physics and chemistry but together with the essential additional elements. The development of life and the evolution process depend crucially on the specific histhory of the system. Living systems learn from experience so they can be thought as complex adaptive systems.[2] In this respect they are different from the (non-adaptive) complex systems mentioned before.

However, there are also some important connections between the two classes. In the past few years relatively simple models, inspired by physics, have been formulated which contain simple elements of adaptation.[3] In addition, it is easy to argue that the development of life requires the presence of a complex enviroment (intermediate between order and disorder) that leads to a sufficient variety of possibility that are necessary for adaptation and succesful evolution. In this perspective non adaptive complexity provides the necessary enviroment for adaptive complexity.

2. Recent Developments in Statistical Physics

In view of the above developments statistical physics is undergoing a profound transformation. The introduction of new ideas, inspired by fractal geometry and scaling, irreversible and non-ergodic dynamical systems leading to self-organization and stochastic processes of various types, leads to a considerable enrichment of the traditional framework and provides efficient methods for characterising and understanding complex systems. The physics of scale-invariant and complex systems is a novel field which includes topics from several disciplines ranging from condensed matter physics to geology, biology, astrophysics and economics.[1] This widespread interdisciplinarity corresponds to the fact that these ideas allow us to look at natural phenomena in a radically new and original way, eventually leading to unifying concepts independent of the detailed structure of systems.

In scale invariant phenomena, events and information spread over a wide range of length and time scales, so that no matter what is the size of the scale considered one always observes surprisingly rich structures. These systems, with very many degrees of freedom, are usually so complex that their large scale behaviour cannot be predicted from the microscopic dynamics. New types of collective behaviour arise and their understanding represents one of the most challenging areas in modern statistical physics.

The interest in this field has been largely due to two factors. First the emerging availability of high powered computers over the past decade has enabled us to readily simulate complex and disordered systems. Second the cross disciplinary mathematical language for describing these phenomena evolving under conditions far from equilibrium has only become available in the past years. The study of critical phenomena in second order transitions introduced the concepts of scaling and power law behavior. Fractal geometry[4] provided the mathematical framework for the extension of these concepts to a vast variety of natural phenomena.

The physics of complex systems, however, turned out to be effectively new with respect to critical phenomena. The theory of equilibrium statistical physics is strongly based on the ergodic hypothesis and scale invariance develops at the critical equilibrium between order and disorder. Reaching this equilibrium requires the fine tuning of various parameters. On the contrary most of the scale free phenomena observed in nature are *self-organized*, in the sense that they spontaneously develop from the generating dynamical process. One is then forced to seek the origin of the scale invariance in nature in the rich domain of nonequilibrium systems and this requires the development of new ideas and methods.

The realization that certain structures exhibit fractal properties does not tell us why this happens but it is crucial to formulate the right questions. The impact of fractals in physics can be assessed along three different lines of increasing complexity:

(a) Fractal geometry merely as a *mathematical framework* which leads to a re-analysis of known data that results in a revamping of long standing points of view. This permits to include into the scientific areas many phenomena characterised by intrinsic irregularities which have been previously neglected because of the lack of an appropriate mathematical. The main examples of this type can be found in the geophysical and astrophysical data and in Sec. 4 we consider one example in detail.

(b) The development of *physical models* for systems that exhibit fractal and Self-Organized Critical (SOC) behaviour. From a mathematical point of view the problems explored are relatively easy to formulate but they consist of iterative systems with many degrees of freedom and irreversible dynamics and lead to extremely complex structures and behaviors. In this respect computer simulations represent an essential method in the physics of complex and scale invariant systems. While the great majority of the theoretical activity is based upon "toy models" which barely resemble real nature, it is important to build a bridge between theory and real experiments and this is another basic task of computer simulations.

(c) The construction of complete *physical theories* that allow us to understand the self-organized origin of fractal structures as well as all the other relevant properties in various physical systems and phenomena. One of the main difficulties in this respect is that the time development is intrinsically irreversible and it cannot be eliminated by some form of the ergodic hypothesis. In equilibrium statistical mechanics it is in fact possible to eliminate the specific dynamical evolution and to directly assign a Boltzmann weight to a given configuration. In the case of self-organized fractal structures this is usually not possible and a full knowledge of the dynamical history is necessary. This implies the development of theoretical concepts of novel type.

3. Regular and Irregular Structures: Scale Invariance

Most of theoretical physics is based on regular (analytical) functions and differential equations. This implies that structures should be essentially smooth and irregularities are treated as single fluctuations or isolated singularities. The study of critical phenomena and the development of the Renormalization

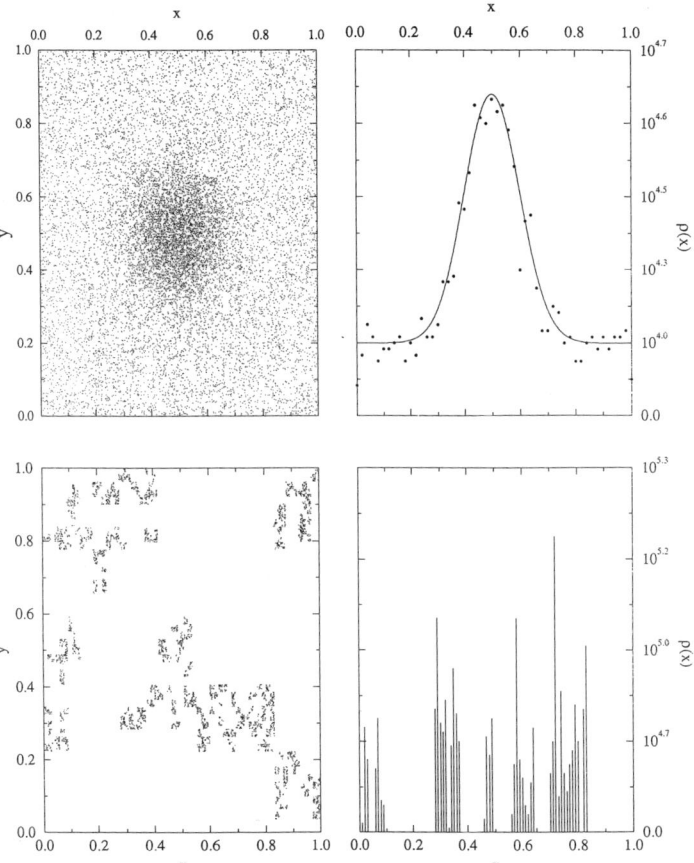

Fig. 1. Example of regular (analytic) and intrinsically irregular structures. Top panels: (Left) A regular structure in a homogenous distribution. (Right) Density profile. In this case the fluctuation corresponds to an enhancement of a factor 3 with respect to the average density. Bottom panels: (Left) Fractal distribution in the two dimensional Euclidean space. (Right) Density profile. In this case the fluctuations are non-analytic and there is no reference value, i.e. the average density. The average density scales as a power law from any occupied point of the structure. This implies that one cannot define the size or the intensity of this fluctuation.

Group (RG) theory in the seventies was therefore a major breakthrough.[5] One could observe and describe phenomena in which intrinsic self-similar irregularities develop at all scales and fluctuations cannot be described in terms of analytical functions. The theoretical methods to describe this situation could not be based on ordinary differential equations because self-similarity implies the absence of analyticity and the familiar mathematical physics becomes inapplicable. In some sense the RG corresponds to the search of a space in which the problem becomes analytical again. This is the space of scale transformations but not the ordinary space in which fluctuations are extremely irregular. For a while this peculiar situation seemed to be restricted to the specific critical point corresponding to the competition between order and disorder. In the past years instead, the development of Fractal Geometry,[4] has allowed us to realize that a large variety of structures in nature are intrinsically irregular and self-similar.

The main difference between the popular fractals like coastlines, mountains, trees, clouds, lightnings etc. and the self-similarity of critical phenomena is that criticality at phase transitions occurs only with an extremely accurate fine tuning of the critical parameters like, for example, the temperature. In the more familiar structures observed in nature, instead the fractal properties are self-organized, they develop spontaneously from the dynamical process. It is probably in view of this important difference that the two fields of critical phenomena and Fractal Geometry have proceeded somewhat independently, at least at the beginning. The fact that we are traditionally accustomed to think in terms of analytical structures has crucial consequences on the type of questions we ask and on the methods we use to answer them.

4. Properties of Fractals: Simple but Subtle

In the late 19th century, various mathematicians studied certain sets and functions that are continuous bu not differentiable: the Cantor set, the Sierpinski gasket, the Julia set etc.[4] A common element of these sets is that they are self-similar, namely they exibit identical structures at different length scales. Their lengths or areas are ill defined and depend on the unit of measure. In the lower part of Fig. 1 we see instead a stochastic fractal in which the property of scale invariance holds on the average.

Then Hausdorff recognized that one could assign a fractional dimension to such sets. In general these concepts remained unnoticed in physics with few exceptions like Perrin, who pointed out that most curves in nature are actually non-differentiable and their length is ill defined. Richardson noted in 1926 that fluid turbulence consists of self-similar eddies and in 1941 Kolmogorov was inspired by this idea of the self-similar cascade to construct his theory in which, however, the energy dissipation is assumed to be homogeneous on space.

Mandelbrot came on the scene in the 1950's and his key contribution was the recognition that self-similarity and non-differentiable curves are pervasive in nature and the so-called pathological functions of fractional dimension constitute the appropriate mathematical language for their description. He codified the scattered mathematical notions of self-similar sets, coined the term "fractal" (from the latin "fractus" or fragmented), and introduced them as a powerful mathematical tool for the scientific study of complex objects and phenomena in the physical world.[4] Mountains, trees, coastlines, lightnings, turbulence in fluids, the distribution of galaxies but also transmission errors in telephone systems, the statistical structure of language, the human lungs and even stock price fluctuations are examples where fractal properties have been clearly detected. Fractal geometry therefore allows us to expand the frontiers of scientific disciplines to include all those phenomena that are characterized by intrinsic, self-similar irregularities. This changes the way we look at nature which now appears much richer and fascinating.

It is useful to consider a few essential properties of fractal structures.[4] Consider an object as a set of points in Euclidean space of dimension d. Use one of these points as the origin and count the number of points $N(R)$ within a distance R. If the object is Euclidean we have $N(R) = R^d$. For a self-similar fractal object, like those shown in Fig. 1 (bottom), one still finds a power law relation between *mass and length*, but in this case $N(R) = R^D$, where D is a non-integer number (the fractal dimension) which, depending on the particular structure, can take any value between 0 and d. From this it follows that the conditional average density of points is $n(R)=N(R)/(R^d)= R^{-(d-D)}$. Thus, fractal objects have a density with power law dependence and hence an *ill-defined average density*. For comparison in a solid with a periodic structure the conditional density consists in a set of delta functions. At very large r, the density of these delta functions approaches a constant, corresponding to the average density of points. For a fractal instead the density is a power law at all scales and *decays continuously* towards zero. This implies that at large scales empty regions will prevail and the structure becomes very sparse. In addition the non-integer power law correlations refer to statistical averages. The corresponding structure is characterized by *scale invariance* that implies *strong fluctuations, clustering* and the presence of *structures and voids of all scales* (see Fig. 1) and the structure in nowhere analytical.

5. The Fractal Universe

In this section we discuss an example in which the concept of Fractal Geometry, used as a mathematical tool, discloses new properties for the large scale strucure of the universe and leads to fascinating and controversial perspectives.

The three-dimensional distribution of galaxies appears quite irregular and it consists of large structures and large voids. In the example shown in Fig. 2 our galaxy is at the center and the empty slice corresponds to the galactic plane in which observations are difficult. Note that the picture is a projection (orthogonal) and this gives a smoothing effect to the eye. If one could rotate this picture as in a video the large structures and large voids would be better defined. Despite these structures the universe is usually believed to be homogeneous at large scale and this property is supposed to be in agreement with the data of Fig. 2.

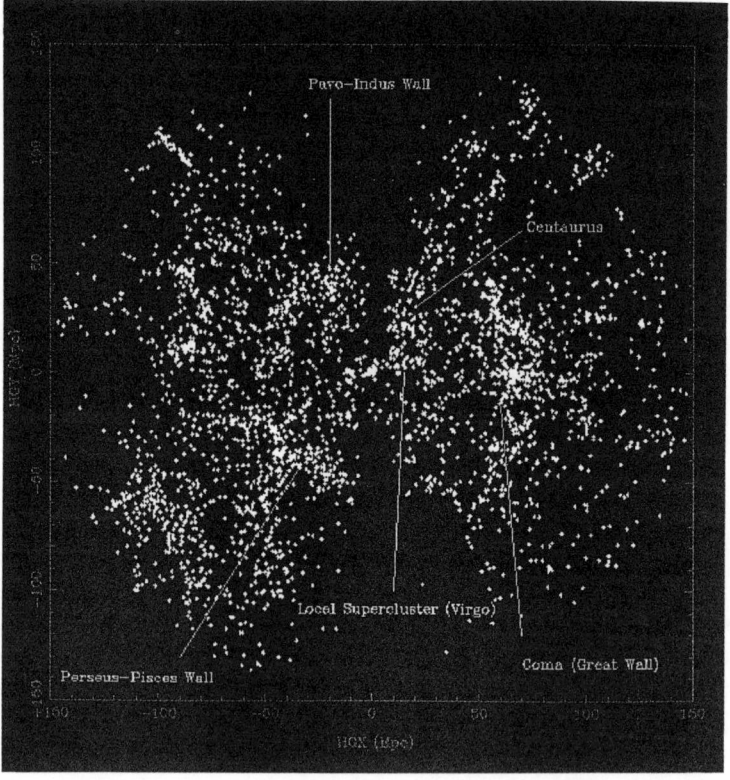

Fig. 2. Three-dimensional distribution of galaxies around our galaxy (central point). Each point corresponds to a galaxy. The area represented covers about one-twentieth of the size of the entire universe. The structure is clumpy and shows clusters and voids at all scales. The empty vertical region corresponds to the galactic plane where observations are limited.

Some years ago we proposed a new approach for the analysis of galaxy and cluster correlations based on the concepts and methods of modern statistical physics. This led to the surprising result that galaxy correlations are fractal and not homogeneous up to the limits of the available catalogues. In the meantime many more redshifts have been measured and we have extended our methods also to the analysis of various other properties.[6–8]

The usual statistical methods, based on the assumption of homogeneity,[9] therefore appear to be inconsistent for all the length scales probed until now. A new, more general, conceptual framework is necessary to identify the real physical properties of these structures and theories should shift from "amplitudes" to "exponents" in the sense discussed in the previous section.

The new analysis shows that all the available data are consistent with each other and show fractal correlations (with dimension $D = 2$) up to the deepest scales probed until now (1000 Mpc).[7,8] In these units, megaparsecs, the radius of the entire universe is about 4000 Mpc, while the size of a single galaxy (a point in our analysis) is about 0.01–0.1 Mpc. The distribution of visible matter in the universe is therefore fractal and not homogeneous. In addition the luminosity distribution is correlated with the space distribution in a specific way characterized by multifractal properties. These facts lead to fascinating conceptual implications about our knowledge of the universe and to a new scenario for the theoretical challenge.

This result has caused a large debate in the field[6] because it is in contrast with the usual assumption of large scale homogeneity which is at the basis of most theories. Actually homogeneity represents much more than a working hypothesis for theory, it is often considered as a paradigm or principle and for some authors it is conceptually absurd even to question it.[9] For other authors instead homogeneity is just the simplest working hypothesis and the idea that nature might actually be more complex is considered as extremely interesting.[10] These two points of view are not so different after all because, if something considered absurd becomes real, then it is indeed very exciting.

The problem is that these concepts touch on directly the so-called Cosmological Principle (CP), which represents one of the landmarks of the field of cosmology. It is quite reasonable to assume that we are not in a very special point of the universe and to consider this as a principle, the CP. The usual mathematical implication of this principle is that the universe must be homogeneous.[9] This reasoning implies the hidden assumption of analyticity that often is not even mentioned. In fact the above reasonable requirement only leads to local isotropy. For an analytical structure this also implies homogeneity.[10] However, if the structure is not analytical, the above reasoning does not hold. For example a fractal structure has local isotropy but not homogeneity. In simple terms this means that all galaxies live in similar enviroments made of structures and voids (statistical isotropy). Therefore a fractal structure satisfies the CP in the sense that all the points are essentially equivalent (no center or special points) but this does not imply that these points are distributed uniformly.[4]

The main results of our new analysis are the following:

— The highly irregular galaxy distributions with large structures and voids strongly point to a new statistical approach in which the existence of a well defined average density is not assumed *a priori* and the possibility of non-analytical properties should be addressed specifically.

— The new approach for the study of galaxy correlations in all the available catalogues shows that their properties are actually compatible with each other and they are statistically valid samples. The severe discrepancies between different catalogues that have led various authors to consider these catalogues as not fair, were due to the inappropriate methods of analysis.

— The correct two point correlation analysis shows well defined fractal correlations up to the present observational limits, from 1 to 1000 Mpc with fractal dimension $D = 2$.

From the theoretical point of view the fact that we have a situation characterized by self-similar structures implies that we should not use concepts which make refence to the average density or related properties. One cannot talk about "small" or "large" amplitudes for a self-similar structure because of the lack of a reference value like the average density. The physics should shift from "amplitudes" towards "exponents" and the methods of modern statistical physics should be adopted. This leads to a new, fascinating situation, that has been uncovered by the introduction of the concepts of self-similarity and Fractal Geometry.

6. Physical Models and Self-Organization

The key question is *how does nature produce fractal structures?* The first physical model that shed light on this question was the *Diffusion Limited Aggregation* (DLA) model of Witten and Sander[11] introduced in 1981. The model was inspired by the observation of growing aggregates that were found to exibit fractal structures. One starts with a seed particle and introduces a new particle at some (large enough) distance R that executes a random walk on a lattice. When the particle reaches a site adjacent to the seed, it is frozen in that position and extends the seed. A new particle is then introduced until it touches the new seed and so on. The iteration of this simple algoritm produces spontaneously structures of great complexity with a fractal dimension $D = 1.7$ (for planar growth). An interesting variant of DLA is the *Cluster-Cluster* aggregation model[12] where one starts with many particles executing random walks that are allowed to aggregate into clusters. Clusters of all sizes continue to execute random walks forming cluster aggregates and so on. Each cluster turns out to be a fractal with dimension that is lower than in the DLA model. In addition the distribution of cluster sizes exibits power-law behavior. The Cluster-Cluster model captures the physics of dust or smoke clouds and colloids[13] as shown in Fig. 3.

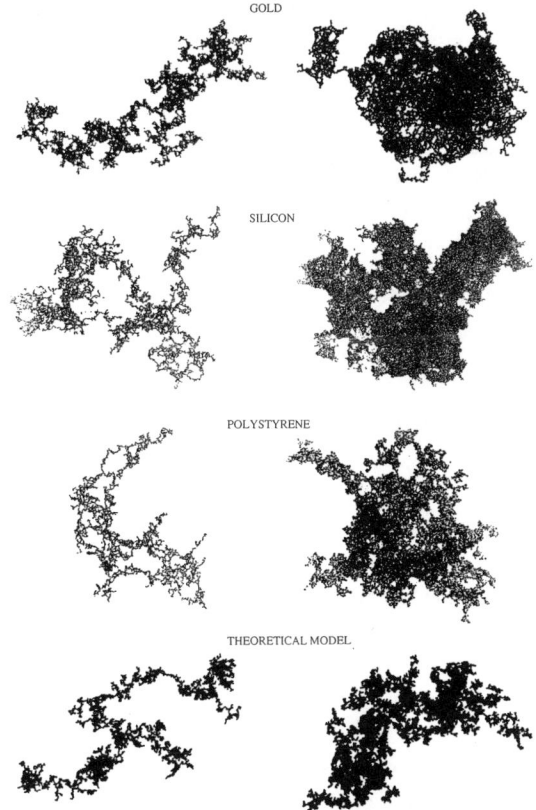

GOLD

SILICON

POLYSTYRENE

THEORETICAL MODEL

Fig. 3. Comparison of various experimental aggregates with those ganerated by a computer simulation of the Cluster-Cluster model (figures at the bottom). The clusters on the left correspond to the irreversible growth limit (low temperature). Those on the right to the reversible case (high temperature).

In 1984 Niemeyer *et al.* introduced the Dielectric Breakdown Model (DBM)[14] inspired by discharges in gases (e.g. lightning). The grown structure is assumed to be eqipotential. From the Laplace equation one can compute the local field for the bonds around the structure. The growth probality is related to the local field. A bond is selected and added to the structure and the process is then iterated. Apart from generalizing the DLA growth process the DBM illustrates the underlying mathematical properties in relation to partial These "Laplacian" fractals are believed to capture the essential fractal properties of a variety of phenomena such as electrochemical deposition, dielectric breakdown, viscous fingering in fluids, the propagation of fractures and various properties of colloids.[12]

The concept of spontaneous generation of complex or critical structures, also called *Self-Organized Criticality* (SOC) has been recently emphasized and investigated in the *sandpile models* introduced by Bak and coworkers.[15] To illustrate the basic ideas of SOC, they introduced a cellular automaton model of sandpiles. The random addition of sand grains drives the system towards a stationary state with a scale free distribution of avalanches. As in the previous fractal growth models, also in this model criticality seems to emerge automatically without the fine tuning of parameters. Because of their conceptual importance SOC ideas have invaded rapidly throughout the sciences, from physics and geophysics to biology and economics, as a prototype mechanism to understand the manifestation of scale invariance and complexity in natural phenomena.

7. The Development of Understanding

The physical models discussed in the previous sections illustrate a number of physical situations that can lead to the generation of fractal structures. Comparison with experimental data suggests that these models capture the essential physics of various phenomena that produce fractal structures in nature. Such models however do not constitute a physical theory and this is the next step of our discussion.

From the theoretical point of view the idea of many authors is that DLA/DBM and the other SOC models pose questions of a new type for which it would be desirable to have a common theoretical scheme.[16] The attempts to use the theoretical concepts developed for critical phenomena like field theory and the RG have been quite problematic for these new phenomena. The basic differences with respect to equilibrium phase transition is that the dynamics is irreversible and self-organized. There is no ergodic principle and it is not possible to assign a Boltzmann weight to a configuration without knowing its entire growth history.

The theoretical effort in this field can be separated into phenomenological or scaling theories and microscopic theories. The first approach has been extensively developed in the past years and it consists in defining consistency relations between the assumed scaling properties of the system. This phenomenological approach is essential in the analysis of computer simulations to identify and extract the relevant essential information.

The microscopic approach consists in a comprehensive understanding of all the SOC and fractal properties of the system directly from the knowledge of the microscopic dynamics. This brings us to the crucial problem of fractal growth. We have seen that most fractal growth models like DLA, DBM, Cluter-Cluster aggregation, the sandpile models are characterized by an intrinsically irreversible dynamics. As a result the statistical weight of a configuration can be defined only with the knowledge of its entire history. Another important difference is that most fractal structures are *self-organized*. One attempt of constructing a physical theory for the self-organized fractals with irreversible dynamics is the Fixed Scale Transformation (FST).[17] This approach allows us to understand the origin of self-organization in fractal growth in terms of an attractive fixed point for the scale invariant dynamics and to compute analytically the fractal dimension. At the moment the FST framework seems to be the only general approach for the broad class of self-organized fractals and related phenomena like the sandpile model.[18]

Despite this progress many important questions remain open. The objective would be to develop these ideas into a general and systematic theoretical framework with microscopic predictive power in relation to fractal growth and SOC properties. For example a crucial issue is the role of universality in fractal and SOC phenomena. In usual critical phenomena the same exponents that define the onset of magnetisation also describe the liquid vapour transition in water. This strong universality appears to be a characteristic of equilibrium systems. Self-organized systems, on the other hand, do not seem to exhibit the same degree of universality as the fractal dimension can be easily altered by relatively simple changes in the growth process. This reduced universality is sometimes viewed as a negative element because one is forced to describe specific systems instead of a single universal model. The truth is probably the opposite. Some theoretical concepts can be considered as general or universal, but the inherent diversity of the various models that have been studied adds another fascinating dimension in the intellectual search. After all the SOC fractal structures we observe in nature are quite various and different from each other. The preliminary knowledge we have at the moment suggests that there are some universal principles but the specific properties depend on the specific process. It is possible that this has to do with the fact that the domain of irreversible phenomena is much broader than that of equilibrium statistics. The definition of the classes and laws for this broader area is certainly one of the main tasks of the theoretical effort in this field.

References

1. C.J.G. Evertsz, H.O. Peitgen and R.F. Voss, (eds.) *Fractal Geometry and Analysis* (World Scientific, 1996).
2. M. Gell-Mann, *The Quark and the Jaguar* (Freeman, New York, 1994).
3. S.A. Kauffmann, *The Origins of Order* (Oxford Univ. Press, New York, 1993).
4. B. Mandelbrot, *The Fractal Geometry of Nature* (Freeman, New York, 1982).
5. K.G. Wilson, *Phys. Rep.* **12** (1974) 75.
6. P.H. Coleman and L. Pietronero, *Phys. Rep.* **231** (1992) 311.
7. L. Pietronero, M. Montuori and F. Sylos Labini, in *Critical Dialogues in Cosmology*, ed. N. Turok (World Scientific, 1997), p. 24.
8. F. Sylos Labini, M. Montuori and L. Pietronero, "Scaling in Galaxy Clustering", *Physics Reports*, in print, 1997.
9. P.E.J. Peebles, *The Large Scale Structure of The Universe* (Princeton Univ. Press, 1980); P.E.J. Peebles, *Principles of physical Cosmology* (Princeton Univ. Press, 1993).
10. S. E. Weinberg, *Gravitation and Cosmology* (Wiley, New York, 1972).
11. T.A. Witten and L.M. Sander, *Phys. Rev. Lett.* **47** (1981) 1400.
12. T. Vicsek, *Fractal Growth Phenomena* (World Scientific, 1992).
13. D. Weitz, *Phys. Rev. Lett.*, **52** (1984) 1433.
14. L. Niemeyer, L. Pietronero and H.J. Wiesmann, *Phys. Rev. Lett.* **52** (1984) 1033.
15. P. Bak, C. Tang and K. Wiesenfeld, *Phys. Rev. Lett.* **59** (1987) 381.
16. L.P. Kadanoff, *Physica* **A163** (1990) 1. See also *Physics Today*, March 1991, p. 9.
17. A. Erzan, L. Pietronero, A. Vespignani, *Rev. Mod. Phys.* **67** (1995) 554.
18. L. Pietronero, A. Vespignani and S. Zapperi, *Phys. Rev. Lett.* **72** (1994) 1690.

Dynamical Systems in Psychology: Linguistic Approaches

William Sulis

Departments of Psychiatry, Psychology, and Computer Science
McMaster University, Hamilton, Ontario, Canada

Major goals for psychoanalysis and psychology are the description, analysis, prediction, and control of behaviour. Natural language has long provided the medium for the formulation of our theoretical understanding of behavior. But with the advent of nonlinear dynamics, a new language has appeared which offers promise to provide a quantitative theory of behaviour. In this paper, some of the limitations of natural and formal languages are discussed. Several approaches to understanding the links between natural and formal languages, as applied to the study of behavior, are discussed. These include symbolic dynamics, Moore's generalized shifts, Crutchfield's ϵ machines, and dynamical automata.

1 Introduction

The ultimate quest of both psychology and psychoanalysis is an understanding of the genesis and meaning of behaviour. Until this century, the main tools applied to this pursuit have been introspection, observation, logical argument and narrative explanation. Quantitative methodologies have been limited mostly to sophisticated forms of classification, counting, and correlation. Although much has been learned through such simple methods, the full power of the quantitative approach has not yet been realized. The advent of nonlinear dynamics and the success of its most visible application, the study of deterministic chaos, has engendered considerable optimism among researchers that a quantitative language is at hand which can be applied to the study of behaviour with as much success as the language of differential equations has achieved in solving the problems of physics.

At the crux of this research is the definition of behaviour itself. We all tend to assume that we understand implicitly what we mean by behaviour. In part, this occurs because, at least within a shared culture, we have learned cultural stereotypes for different interpersonal scenarios which provide norms for the classification of behaviour. That these norms are to a large extent arbitrary is readily apparent to anyone who has engaged in the practice of transcultural psychiatry. The arbitrariness of behavior constitutes a deep impediment to the development of any theory concerning its organization and generation. Paulus and Geyer have studied this question in some detail within the context of their work on locomotor behaviour in rodents. They write (Paulus & Geyer, 1993):

Behavioral patterns can be defined as particular combinations of event sequences observed within a collection of behavioral events. However, the definition of a distinguishable, discrete behavioral event poses a difficult problem. Frequently, this problem has been circumvented by a-priori definitions of categories of behavioral events and subsequent quantification of behavioral patterns by virtue of the frequency distribution of these categories. First, the idiosyncratic nature of the definition of a specific behavioral category depends on the specific measuring device and testing environment. Therefore, behavioral responses are difficult to compare in different paradigms measuring motor activity. Second, different behavioral categories may be highly correlated, resulting in an explosion of the number of different measures without necessarily providing additional information about the behavior of the animal. Third, some manipulations, such as the prototypical psychostimulant d- amphetamine, lead to a fragmentation of behavioral responses making it difficult to assess whether a specific behavioral response has been completed. In addition, the use of predefined behavioral categories precludes quantitative assessment of the transition from normal behavioral activity to fragmented or shortened behavioral sequences. Fourth, a-priori defined behavioral categories may not correspond to distinct neurochemical processes or neuroanatomical systems, complicating attempts to determine the neurobiological substrates of behavioral effects. Fifth, behavioral categories are not easily monitored with the temporal resolution that is necessary to detect subtle changes in the response topography of the animal. In conclusion, the assessment of patterns of motor activity obtained from behavioral observable should be relatively independent of the measuring resolution and the measuring device, should not impose an a-priori distinction of behavioral categories, should yield uncorrelated information about different aspects of motor behavior, should be obtained with a high temporal frequency, and should be able to assess the fragmentation and disintegration of behavioral sequences.

Thus we have a deep and fundamental problem determining what a behavioral act is and how to properly partition an animal's activity into such acts.

A more fundamental problem exists, however. Golani (1992) writes:

> In a paper on "gestalt perception as a source of scientific knowledge", Lorenz explained why gestalt perception is an indispensible step in the establishment of behavioral homologies; he suggested that this process is basically intuitive and subconscious and therefore cannot be taught. What Lorenz did not consider was the role that language- any language- plays, in creating a disposition for perception and thought and in organizing experience. The gestalt perception of movement is programmed by the language used by the observer, and the terms and words of a language are the vehicles and tools of perception and thought. Everyday terms obscure both what is common and what is different in the structure of behaviour.

Thus we also face a deep problem with respect to determining what language should be used to describe behavioral acts. It is to that question that we now turn our attention.

2 Language Natural and Unnatural

Natural language has long provided us with a powerful heuristic for the description and analysis of behaviour. Natural language provides a fundamental mode of communication, whether external or internal. Linguistic skills are acquired without conscious effort, leaving us with a feeling of intimate connection to language. Natural language appears inextricably linked to conscious experience. Unfortunately, this deep and facile association with natural language leads us to assume that natural language provides us with a unique and privileged capacity for understanding. We confuse the medium and the message.

No language provides us with a direct experience of the world. Instead, language provides us with a set of referents, and a means of relating them, through which we attempt to create an empathic relationship with external reality. All language is metaphorical when applied to understanding this reality. At best, we create a stream of internal experiences which resonate with salient aspects of the external experience which we are attempting to comprehend. The faithfulness of such resonance, and the degree to which the association can be generalized beyond these points of resonance, is a matter of great concern to scholars and scientists alike. All too often our faith in our own intuitive understanding causes us to attribute a far greater fidelity to our metaphors than is often justified by the available facts.

All language is symbolic. This tautology leads inexorably to a fundamental pragmatic and philosophical problem - the symbol grounding problem. A symbol, to be useful, must be grounded in external reality. The grounding of natural language symbols occurs within each individual as a result of the accumulation of life experience through learning, and within each culture through consensual validation arising from social discourse. Biological reality excludes the direct sharing of experience between individuals and as a result, the grounding of natural language symbols within each individual is both unique, and private. This places a severe limitation upon the public use of such language as a vehicle for understanding. Culture provides a body of consensual referents, but these evolve over time, locking the grounding of such cultural symbols to specific times, and specific places.

Language provides only a metaphorical understanding of reality. This understanding is achieved through the construction of a chain of metaphors, leading from the object to some internal experience of the observer. Psychologically then, the observer projects this inner experience back along the chain to the original object. This produces within the observer a sense of an empathic link with the object. But such a link is unique to each individual. Moreover the depth of the understanding which is achieved through this chain is dependent upon the resonance between each link in this chain and the corresponding reality. Opportunities for confusion, error, and delusion abound.

Only the external, public features of language lend themselves to consensual validation and the establishment of consensual consistency. However there is seldom an exact consensus about what constitutes a properly formed expression in a natural language, even from the standpoint of grammar and available vocabulary. Logic notwithstanding, there is seldom consensus about how to derive valid expressions from previously established expressions. There are conventions, but no rules. Natural languages lack consistent structure. Irregularities abound. There is also a lack of internal grounding. A word may play several different roles depending upon context. And the context often requires reference to features outside of the language itself. In addition, most natural languages lack quantification, restricting their usage in quantitative science.

This century has seen a rapid growth in the development of unnatural languages. The most widely recognized is mathematics. It has had a long and illustrious history. And it has undergone unprecedented growth during this century. But this century has also witnessed an explosion of artificial languages as a result of the introduction of the digital computer. Operating systems, programming languages, simulation environments and Internet protocols all constitute languages in the formal sense of the term. The field of mathematical linguistics arose in an attempt to understand the relationship between these formal languages and the objects which they purport to represent. Although these artificial languages are, in themselves, still too crude to use as vehicles for understanding. mathematical linguistics provides a formal approach to understanding at least the structure of language.

Unlike other languages, mathematics possesses a dual nature. It is a language which can be used to convey information. But it is also a subject in its own right, and can be used self referentially to describe itself. This provides it with both power, and limitation, painfully demonstrated through Godel's incompleteness theorem. Mathematics is internally grounded by through its internal consistency, which can be established on the basis of logic alone. Consensus about the meaning of mathematical expressions is public, through a shared culture and education, and possesses a high degree of consistency, which transcends traditional cultural barriers. It possesses a logical structure. Expressions can be readily evaluated for their structural correctness, even if, as a result of the incompleteness theorem, it may not always be possible to determine their truth. Procedures exist for generating true statements from true statements. Mathematics is also fundamentally quantitative in nature and so readily lends itself, almost by definition, to quantitative science. For these reasons, mathematics serves as a preeminent language for description, prediction, and control.

In the final analysis though, the application of mathematical formalism to the understanding of real world problems remains as metaphorical as for natural language. Mathematics cannot escape the same external symbol grounding problem as does natural language.

3 Dynamical Systems Theory

In recent years there has been a growing number of books [1,6,11,17] which have suggested that nonlinear dynamical systems theory may provide a quantitative language suitable for the description, understanding, prediction and control of psychological and sociological processes. All psychological phenomena arise within situated, embodied agents and occur within finite time intervals. It is a fundamental assumption of modern psychology and psychoanalysis that psychological phenomena are a high level expression of the temporal evolution of psychophysical and neurobiological states. Thus psychological systems are, first, and foremost, dynamical systems. So it seems natural that the formal study of dynamical systems be applied to the study of psychological processes.

In formal terms, a dynamical system consists of a set X of states together with some means of specifying how these states relate to one another over the course of time. The traditional formulation is by means either of differential equations such as the Schrodinger equation if time is continuous:

$$-(\hbar/i)\partial\psi/\partial t = -(\hbar^2/2m)\partial^2\psi/\partial x^2 + 1/2kx^2\psi$$

or, if time is discrete, by an iterated function such as the logistic map:

$$f_\mu = \mu x(1 - x)$$

Nonlinear dynamics has already yielded insights into the generation and selection of behaviour. The neural mechanisms responsible for the generation of specific behavioral acts have been studied in diverse settings, for example the role of coupled oscillators in lamprey (Kopell & Ermentrout, 1988) or of dynamical symmetries in gait (Collins & Stewart, 1993). Models directed towards understanding the organization of specific acts into coherent behavioral patterns are still scarce. The study of the organization of behaviour requires an understanding of the generation of spatiotemporal patterns, something poorly understood even within the hard sciences.

There are problems, however, in applying nonlinear dynamical systems theory to psychology. Nonlinear dynamical systems theory is fundamentally a deterministic theory. As such, it is incapable of dealing with the myriad forms of noise and unreliability which pervade living nervous systems. The attempt to understand this noise in terms of deterministic chaos only obscures the issue, since replacing noise with ad hoc chaotic dynamics introduces an unnecessary burden of unobservable parameters, and does not render the overall system any more tractable to formal analysis. There is of course, a rich theory devoted to the study of stochastic processes, but this theory fails to deal with the additional difficulties described below.

All dynamical systems theories, stochastic or deterministic, posit the existence of infinitesimal states and instantaneous observation times. Behavior, however, has extension both in time and in space, and is generated by agents whose their material extension is a fundamental component of those processes which are responsible for the generation of the behaviour. Moreover, dynamical systems theories are primarily concerned with the long time behaviour of systems. They are virtually incapable of dealing with transients. Yet all behavior is transient. Theories of behaviour must eventually address the problems of transience and of extensionality. Moreover, virtually all dynamical theories presume that the underlying dynamics is differentiable. This is far from evident as a feature of neural dynamics. Both continuous and discrete processes occur.

Most dynamical systems theories concern systems which are isolated from their environments and which are operating autonomously. But the environment plays a fundamental role in the generation of behaviour. The standard artifice of incorporating the environment into the equations, producing a new autonomous system does not adequately address this issue since it is the relationship between the environment and the behaving system which is of most interest to psychologists, and of particular relevance to psychoanalysts and psychotherapists.

Finally, the term behaviour applies to coherent sets of action, not specific actions. Dynamical systems theories, though frequently concerned with questions of stability, are seldom concerned with questions of coherence.

In spite of these reservations, nonlinear dynamical systems theory does provide a conceptual, and frequently mechanistic foundation upon which to found a theory of psychology, and so it behooves us to at least take its concepts and formulation into consideration when attempting to forge an understanding of behaviour.

We thus have two languages which we can apply to the description and understanding of psychological behaviour: natural language, and dynamical systems theory. Both possess strengths, and both have weaknesses which render neither likely to provide a final answer. Yet one wonders whether it might not be possible formulate a new language, resting on their strengths and avoiding their weaknesses.

It seems to me that the challenge which faces us is to forge an understanding of the following two problems:

1. What is the relationship between dynamical systems theory and natural language, or, more formally, what is the relationship between dynamical systems theoretic and linguistic theoretic descriptions of dynamical and complex systems behaviour?

2. Which dynamical properties are necessary and sufficient to give rise to linguistic behaviour and structure in the observable behaviour of a system? By this I mean elucidating which properties are necessary and sufficient so that the behaviour of the system possesses an intrinsic linguistic description. From this study one hopes to understand what characteristics are required in order for a dynamical system to be a generator and manipulator of language.

To this end, let us explore some research which has been directed towards answering these questions.

4   Symbolic Dynamics

Let me begin by considering the question of the relationship between natural language and dynamical systems approaches. First I will describe a typical psychobiological experiment and its natural language formulation.

Consider a typical experiment in rodent locomotion. The animal is placed in a large open field, usually a square box, about 2 metres on a side. The box is subdivided into smaller subregions which are individually labelled. The animal is then prepared according to whatever paradigm is under study, and then placed in the box and allowed to freely behave within its walls. The path that the animal takes is represented as a series of subregion labels, corresponding to the sequence of subregions physically visited. Generally additional information is appended to this subsequence such as the time of entry into the subregion, the time spent there, the time of departure, plus perhaps a recording of other kinds of behaviours, such as rearing, grooming, or mouthing. This sequence provides a coarse grained description of the actual path which the animal followed during the period of observation.

Symbolic dynamics[4] performs a similar coarse graining for dynamical systems. One begins with a finite alphabet, $A = \{a_0, a_1, \ldots, a_n\}$ and constructs the set of all bi-infinite sequences over this alphabet, $L(A)$. In other words, one considers sequences of the form

$$\ldots s_{-3} s_{-2} s_{-1} . s_0 s_1 s_2 s_3 \ldots$$

An iterative dynamical system is then constructed having $L(A)$ as its phase space together with a dynamic defined by the shift map σ which is defined as

$$\sigma \ldots s_{-3} s_{-2} s_{-1} . s_0 s_1 s_2 s_3 \ldots = \ldots s_{-3} s_{-2} s_{-1} s_0 . s_1 s_2 s_3 \ldots$$

In other words, σ shifts the sequence one place to the left.

Given an iterative dynamical system (X, f), with f invertible, partition the phase space X into a finite set of regions and label them by the alphabet A. Next, beginning with some initial point $x \in X$, construct the trajectory of x under f, in other words, the sequence

$$\ldots f^{-3}x, f^{-2}x, f^{-1}x . x, fx, f^2x, f^3x, f^4x, \ldots$$

Now, each point will lie in one of the subregions defined by the partitioning of the phase space, and so replace each point in the above sequence by the label corresponding to the subregion which contains it, thereby obtaining a sequence $g(x)$ in $L(A)$. If the map f is not invertible, one will obtain only the infinite subsequence corresponding to all terms to the right of the dot.

Now note that the image of fx is just the image of x, but shifted to the left one place. That is, $g(fx) = \sigma g(x)$. Thus the mapping g produces a symbolic representation of the phase space, where the dynamics has been simplified to that of the shift map, σ. In general, the mapping g will not take X to all of $L(A)$, nor will it be one to one, so that the image of each point in the phase need not be unique. However, under certain conditions, a partition (for example a Markov partition) may be constructed in which this mapping is one-one. In that case, we say that the original dynamical system is conjugate, via g, to the shift map.

This happens in the case of the logistic map, $f = \mu x(1-x)$. Here the Markov partition is defined as $0 = [0, 1/2]$ and $1 = (1/2, 1]$. For $\mu = 4$, it can be shown that this partition results in an isomorphism between $[0, 1]$ and $L^*(0, 1)$, the set of all infinite sequences over $0, 1$. From this result, it is an almost trivial exercise to show that the original logistic mapping is chaotic[4].

Moreover, using this symbolic representation, it is possible to show that the dynamics of the logistic map is equivalent to that of tossing a fair coin infinitely often, again demonstrating the close connection between chaos and our notions of randomness[10].

The power of the symbolic dynamical method lies in the inherent universality of symbol systems. Dynamical systems possess a host of attributes, both natural and formal, which render each unique both in the real world and in their formal descriptions. The use of symbolic descriptions cuts across all of these differences to find what is common and universal among them. However the power of symbolic dynamics depends upon the nature of the mapping g. It is most evident in the case of the logistic mapping, where g is an isomorphism. It is less evident say, for $\mu = 3.9$, where the mapping is only onto a subset of $L^*(0, 1)$. To use the symbolic method here requires determining what this subset of sequences looks like, and that is far from trivial. In the case of $\mu = 3.9$, it takes a special form, giving rise to what is known as a subshift of finite type. But the situation is much more complex that than.

Moore[8] has generalized the symbolic dynamics approach to include a much larger class of functions. In so doing, he has made clear some of the profound issues at the heart of this method and a deeper appreciation of its limitations.

Let Z denote the set of integers. Let f be some map $L(A) \xrightarrow{f} Z$, where as before, A is some finite alphabet. Moore states that f has a finite domain of dependence if the set of integers i, such that $f(a)$ depends upon a_i, is finite.

Now Moore defines a generalized shift as follows:

A generalized shift map $\Phi : L(A) \to L(A)$ is defined by

$$\Phi : a \to \sigma^{F(a)}(a \oplus G(a))$$

where

1. $a \in L(A)$
2. $F : L(A) \to L(A)$ has finite domain of dependence
3. $G : L(A) \to L(A \cup \{\emptyset\})$ has finite domain of dependence. $\emptyset$ here is simply used as a place holding symbol.
4. Every sequence in the image of G has a value of $\emptyset$ except for a finite number of cells (not necessarily the domain of dependence)
5. For every $a \in L(A)$ and $g \in L(A \cup \{\emptyset\})$, $a \oplus g$ is defined as follows:

$$(a \oplus g)_i = \left\{ \begin{array}{ll} g_i & \text{if} \quad g_i \neq \emptyset \\ a_i & \text{if} \quad g_i = \emptyset \end{array} \right.$$

Depending upon a finite number of cells in the sequence, the action of the generalized shift is to change some cells in the sequence, according to G, and to then shift left or right by an amount determined by F.

At first glance, it does not appear that this generalization provides anything new or interesting as far as symbolic dynamics is concerned. But Moore was able to prove that every generalized shift can be simulated by a Turing machine, and, moreover, that every for every Turing machine, there exists a generalized shift which is conjugate to it. This is an extremely important result since it places fundamental limitations upon our ability to predict the behaviour of dynamical systems.

Turing machines generate two classes of data, or two kinds of languages: recursive, and recursively enumerable. Recursive languages are those whose membership can be determined exactly by a Turing machine. By this I mean it is possible to find a Turing machine which will always halt and thereby determine whether a given string of letters (or a given set of data) is an element of the language, or not. The famed Church-Turing thesis asserts that any possible computation can be carried out by a Turing machine, so that only languages which are recursive can, in principle, be known exactly. Recursively enumerable languages are those for which a Turing machine will halt, but only when given a string of the language. If a string from outside of the language is presented to the Turing machine, it will not halt. The problem is that it is impossible to know in advance how long it will take the Turing machine to halt. The only way to know if a string is within this recursively enumerable language is to give it to the Turing machine and wait, possibly for a infinite time to see it if halts. To make matters worse, there exist so called non recursively enumerable languages, in fact an infinite hierarchy of these, and I leave it to the reader's imagination to try to conceive what these would be like.

Perhaps the most significant result of Moore as far as description, control and prediction is concerned is that the finite time behaviour of a generalized shift is recursive, but the long time behaviour is recursively enumerable. Thus any dynamical system which is conjugate to a generalized shift shares the same property. This result is even more extreme than simple chaos. In a chaotic system, exact, infinitely detailed knowledge of the initial state of a system will allow its exact long time behaviour to be predicted. However, if the dynamical system is conjugate to a generalized shift, then even exact, infinitely detailed knowledge of the initial state will still be inadequate to predict the long time behaviour. The only way in which one can know the long term behaviour of such a system is to simulate it, and wait.

Our knowledge of sociodynamics is still too primitive determine whether such dynamics might be conjugate to normal shifts, generalized shifts, or something even more extreme. One must bear in mind though that any dynamic more extreme than a generalized shift will have virtually no predictability at all.

There are other problems though in applying symbolic dynamical methods to psychological problems. The decomposition of the open field into sites appears to provide a symbolic description of the kind used in symbolic dynamics. This is not exactly the case however and care must be taken when applying these ideas to real world situations. The first consideration is a subtle one. Let us consider what is meant by a datum from one such experiment. A datum consists of a pair of values: the site in which the animal carries out a very specific behaviour- either stopping or leaving, and the time at which this behaviour occurred. There is no specification of what happens between two such events. For example suppose that the animal exits site 25 at time 2:30.25 and stops at site 35 at time 2:31.55. Any trajectory beginning in site 25 and ending in site 35, of duration 1.30 and without stops is possible. The datum (25, 2:30.25)(35,2:31.55) represents not a single trajectory or a single site but instead represents a set of trajectories defined by these initial and final conditions. Contrast this with the symbolic dynamics case in which each symbol represents a set of individual states. Enough information is conveyed by the infinite sequence of symbols to enable one to recover the structure of the original space. This is not possible here. There is far too much uncertainty associated with each site to make this possible. Thus it is not at all clear that any information about the underlying dynamics of the system will be preserved by this symbolic representation.

The second consideration involves time. The locomotion of the rat occurs continuously in real time. The representation of this path involves a discretization of time. The intervals between each event in the representation are implicitly defined and nonuniform in size. This corresponds to neither a straightforward discrete iterated function system, nor to a symbolic dynamics system in which time is discrete but uniform. A method does exist for reducing a continuous dynamical system into a discrete dynamical system with an intrinsic time. This is the method of Poincare sections[4]. A Poincare section of a continuous dynamical system is obtained by selecting a slice through the phase space of the system and a specific trajectory of the system and then noting the points of intersection between the trajectory and the slice. Ordering these intersection points according to their time of occurrence gives rise to a discrete sequence of points. It can be shown that many properties of the original trajectory are reflected in related properties of this sequence. Thus one obtains a discretization of a continuous system which nevertheless still conveys information about the nature of that continuous system. Unfortunately there are no general techniques available for constructing such Poincare surfaces. Thus there is no guarantee that a decomposition which is artificially imposed upon the open field through the partition possesses the necessary properties of a Poincare section.

Thus although the data sequence obtained from these experiments possesses features suggestive of symbolic dynamics or Poincare sections, it is not a priori obvious that such is actually the case and caution must be used in interpreting the results obtained from such data in dynamical terms. Clearly, a more detailed study of the relationship between symbolic dynamical methods and natural language methods is in order.

5 Crutchfield's ϵ machines

So let us turn to the second question, the attempt to understand which dynamical properties give rise to linguistic behaviour in the system. There appear to be two main approaches to this question in the literature. The first, development by Crutchfield[2], will be discussed here. The second, the dynamical automaton approach, is discussed in the next section.

The basic idea of Crutchfield is to take an observed data stream and to construct a finite state automaton which is capable of reproducing the data. In constructing this machine, a set of states are generated which reflect patterns of dependency in the original data stream. Crutchfield calls this machine, an ϵ-machine, where ϵ refers to the size of the coarse grained partition used to observe the original phase space. This partitioning of the space is equivalent to that described above for symbolic dynamics and the locomotor experiments but, as in the latter case, need not produce a Markov partitioning of the phase space. Thus it comes closer to the situation experienced in real experiments. The data is then passed through a fuzzy measurement device, the details of which are unimportant here, and finally through a binary encoder, to produce a two valued sequence.

The machine is constructed based upon the idea that a state corresponds to a set of subsequences of the original sequence which render the future independent of the past.[3] One constructs these states as follows. Fix a value D, and represent all length D subsequences observed in the original sequence as paths in a depth D binary parse tree. Parse this tree, searching for morphs, the sets of subsequences which follow from a given state. The morphs are found by associating them with distinct depth D/2 subtrees found in the parse tree down to depth D/2. The state to state transitions are found by looking at how each state's associated subtrees map into one another on the parse tree. The result is a stochastic finite automaton, where the transition frequencies are determined by the number of occurrences of a given state from the particular state over the tree.

In parsing the tree, one defines an equivalence relation, declaring two vertices equivalent iff their associated subtrees are identical. That is, $n \sim n'$ iff $T_n^L = T_{n'}^L$. From this we get a set of equivalence classes on the vertex set. These equivalence classes define the states of the automaton.

A second graph is now created, whose vertices consist of these equivalence classes. These new vertices are connected by an edge $v_k \overset{s}{\to} v_l$ if a transition exists in **T** between nodes in the equivalence classes, that is, if there exists a link $n \overset{s}{\to} n'$ where $n \in v_k$ and $n' \in v_l$.

Crutchfield is able to use this modeling strategy quite effectively in developing an ϵ hierarchy for a given dynamical system, and for developing quantitative tools for the analysis of complex systems. In particular he has studied unimodal maps on the unit interval having negative Schwartzian derivatives. The prototypical example of such a map in the logistic map. He has studied the period doubling cascade leading to chaotic activity. One major result of this work has been a linguistic description of this phase transition which occurs at the transition point between periodicity and chaos. In order to use the ϵ machine approach, it is necessary to study the growth in the size of the minimal automaton which is generated from the time series, as a function of the length **D** of the binary parse tree used for the reconstruction. He has demonstrated that for the periodic region, the minimal automaton remains finite in size as **D** increases. Thus the minimal machine which reconstructs a time series within the periodic regime will have a finite number of vertices. However, at the critical point of the transition, the minimal machine size diverges to infinity. As a result, the minimal machine which reconstructs the time series at the critical point has an infinite number of states. If we consider the language generated by these automata, then another interesting phenomenon appears. The languages of the automata in the periodic regime are all regular languages, generated by deterministic finite automata. At the critical point, however, the language generated appears to be an indexed context free language, which is two steps up the Chomsky language hierarchy. Such languages are generated by nested stack automata. It is interesting to note that even though the minimal automaton generating the time series at the critical point has an infinite number of states, the nested stack automaton which generates the associated indexed context free language has a finite description.

Although the ϵ machine approach is clearly a powerful tool for analyzing simple complex systems, it does not adequately address the question of what allows linguistic structure to arise naturally within a dynamical system. The machine is constructed externally to the original system. It describes the system, but need not preserve isomorphically the dynamics of the system. The linguistic structure so obtained is not inherent in the dynamics of the system, but instead is imposed upon it externally. In addition, like standard symbolic dynamics, it deals with single instantaneous states, and does not address the transience issues. In order to deal with these, we turn to dynamical automata.

6 Dynamical Automata

Dynamical automata were first introduced in Sulis.[13,14] They were specifically developed into order to study the transient behavior of dynamical systems, and in particular to study the relationship between environmental transients and system transients, which, as I have argued elsewhere[13] provides an initial model of the classical psychological situation. The concept of a dynamical automata has recently been generalized to provide a foundation for a formal theory of collective intelligence.[18] Some of the results presented below appear without proof Sulis.[14]

We begin by defining the notion of a pattern space, which, intuitively, can be likened to the sequence of frames on a strip of film. The standard notion of a state would correspond to the color at a specific site on a specific frame of film. We begin with a series of definitions.

Definitions:

1. A semigroup is a set X together with an operation $*$ such that for any elements $x, y, z \in X$ we have, $x * y \in X$ and $x * (y * z) = (x * y) * z$. An identity is an element e such that $e * x = x * e = x$.

2. A semiring consists of a set X together with two operations, $+, *$ and two elements i, e such that X is a semigroup with identity i under $+$ and a semigroup with identity e under $*$. In addition we require that $*$ distribute over $+$, that is, $x * (y+z) = (x*y) + (x*z)$ and $(y + z) * x = (y * x) + (z * x)$.

3. A quasisemiring consists of a semiring together with a special element $\emptyset$, the null element such that $\emptyset$ is an identity for both $*$ and $*$. $\emptyset$ acts like a place holder.

4. A quasialgebra consists of a pair X, R of quasisemirings together with an operation of R on X such that for $r, s \in R, x, y, z \in X$ we have $x * (ry+sz) = r(x*y) + s(x*z)$ and $(ry+sz) * x = (r(y*x) + s(z*x))$.

We may now define a pattern space as a triple (P, X, T) where:

1. T is a totally semigroup with identity 0

2. X is a quasisemiring with identities i, e and null $\emptyset$.

3. P consists of a set F, operations $*, +, |$, an operation $\quad$ of X on F, and a mapping τ of F into T. For each $s \in T$, let $F(s)$ denote the set of elements of F which are mapped to s by τ. Then

   (a) each $F(s)$ is a quasisemiring under $*, +$

   (b) each $F(s)$ is a quasialgebra over X

   (c) F is a noncommutative semigroup over $|$

   (d) $|$ distributes over $*, +$

   (e) $F(s)|F(w) = F(s + w)$

As an example, consider the totally ordered semigroup T consisting of the positive real numbers. For $b \in T$, $[0, b) = \{x \mid 0 \le x < b\}$. Also $[0, 0) = \emptyset$ and $[0, \infty) = T$. Let $T^* = \{[0, b) \mid b \in T \cup \{\infty\}\}$. For some set X, define $T^*(X) = \{f \mid p \stackrel{f}{\mapsto} X$ for $p \in T^*$, f bounded continuous$\}$. That is, $T^*(X)$ consists of bounded continuous functions from elements of T^*, that is, intervals on T, to X. Given $f, g \in T^*(X)$, we may define an operation of concatenation fg as follows: Suppose that $[0, b) \stackrel{f}{\mapsto} X$ and $[0, d) \stackrel{g}{\mapsto} X$. Then $[0, b)[0, d) \stackrel{fg}{\mapsto} X$ and

$$fg(x) = \begin{cases} f(x) & \text{if } x \in [0, b) \\ g(x - b) & \text{if } x \ge b \end{cases}$$

On $[0, 0)$ we define the empty function and denote it as 0.

Under these conditions, $T^*(X)$ is a pattern space.

In order to define a dynamical structure, consider a subsemigroup G of the semigroup of transformations on P. Just as P is graded by T, so too is our semigroup of dynamical transformations. Let us restrict our attention to transformations of the following form. Let G be the union of sets of the form $G(t)$, for $t \in T$ where $G(t) = \{g \mid g : F(s) \to F(s + t)\}$ for all s, and $G(t)h = hh'$ for some $h' \in F(t)$. These are termed dynamical operators.

As an example, consider $T^*(X)$ and a continuous dynamical system on X. For any interval $[0, a)$, and function f defined upon this interval, we may extend f to the closed interval $[0, a]$. Let the value at a be c. Then, using c as an initial condition, we may determine the trajectory of the system over the interval $[0, b)$. This results in a map f' on the interval $[0, b)$. Then we define a dynamical operator $g \in G(b)$ by setting $gf = ff'$. Thus any dynamical system has a representation as a dynamical automaton.

There is a second class of operators which are of importance for such systems. These are termed observation operators and consist of a subsemigroup H of transformations on P which is the union of sets of the form $H(t)$ where $H(t) = \{h | h : P \to F(t)\}$. Observation operators $h \in H(t)$ may be forward, i.e. there exists a dynamical operator $g \in G(t)$ such that $g = fh(f)$ for $f \in F(s)$, or backward, meaning that $h(f)f' = f$ if $t < s$.

The prefix operator $P(s)$ and suffix operator $S(s)$ are the unique operators such that, for $s + s' = w$, $s, s', w > 0, x \in F(w)$ implies $x = P(s)xS(s')x$ and $P(s)F(w) = F(w) = S(s)F(w)$ if $s > w > 0$.

Each transformation $G(s)$ has a corresponding forward projection operator $SG(s)$ defined by $SG(s) = S(s)G(s)$.

A dynamical automaton is a pair (P, G) where P is a pattern space and G a semigroup of dynamical operators.

We are particularly interested in dynamical automata which are being driven by some external input. There are several ways in which this notion can be formulated, but the simplest is to merely associate each input e with a corresponding dynamical operator which gives the effect of the input on the subsequent behaviour of the system.

Definition: A dynamical automaton is a 5-tuple (P, G, E, m, Δ) where (P, G) is a dynamical automaton, E a pattern space, $m : E \to G$ and $\Delta : P \times E \to P$ is defined by $\Delta(\psi, \eta) = m(\eta)\psi$.

Definition : Let (P, G, E, m, Δ) be a dynamical automaton. Two patterns $\psi, \psi' \in P$ are dynamically equivalent if for any $\rho \in E$ we have $\Delta(\psi, \rho) = \psi\bar{\psi}$ and $\Delta(\psi', \rho) = \psi'\bar{\psi}$.

Lemma: If W is a set of dynamically equivalent states, then $\Delta(W, \bar{E}) = W\bar{\Delta}(W, \bar{E})$.

Definition: Let (P, G, E, m, Δ) be a dynamical automaton with input. An information event is a pair (Q, R) where $Q \subset E, R \subset P$. Q is the question, R the response. An instance of an information process $\mathcal{P}$ is finite sequence of information events. Given a set $M = \{(Q_i, R_i)\}$ of information events, let $\mathcal{M}$ denote the set of all instances formed from elements of M. $\mathcal{M}$ is just the free monoid over M under concatenation. An information process $\mathcal{P}$ is a subset of $\mathcal{M}$. Therefore $\mathcal{P}$ is a language, an instance is a word and an event is a letter.

Definition: Let (P, G, E, m, Δ) be a dynamical automaton with input. Let (Q, R) be an information event.

1. A subset $\bar{S} \subset S$ is said to support (Q, R) if

(a) For any $\psi \in \bar{S}$, $\bar{\Delta}(\psi, Q) \subset R$

(b) $\Delta(\bar{S}, Q)$ has non empty interior.

2. Let $W = \{(Q_i, R_i) \mid i \in I\}$ be an instance of an information process, and $\bar{E} \subset E$. Then $\mathcal{A}$ supports W in $\bar{E}$ if there exists a subset $\bar{S} \subset S$ such that $\bar{S}$ supports (Q_1, R_1) and for all $j > 1$ we have that $\Delta(\bar{S}, Q_1\bar{E} \cdots Q_{j-1}\bar{E})$ supports (Q_j, R_j).

3. Let $\mathcal{P}$ be an information process. We say that $\mathcal{A}$ processes $\mathcal{P}$ effectively if there exists a subset $\bar{S} \subset S$ such that $\bar{S}$ supports W for any instance $W \in \mathcal{P}$.

Most systems of interest will, of course, be dependent upon their own unique history. A full theory of such systems is beyond reach at present. So attention will be restricted to those systems whose past dependency extends for only a fixed, finite, time into the past.

Definition: Let $(P.G, E, m, \Delta)$ be a dynamical automaton with input. It is termed t-local if $\Delta(\psi\sigma, E) = \psi\Delta(\sigma, E)$ for every $\psi, \sigma \in S$ such that $\tau(\sigma) \geq t$.

Lemma: (P, G, E, m, Δ) be a dynamical automaton with input.. Assume that it is t-local and that W is a set of dynamically equivalent states such that $\tau(\psi) \geq t$ for all $\psi \in W$. Then for any $H \subset SW$ and any $F \subset E$ we have $\Delta(H, E) = H\bar{\Delta}(H, F)$.

The main result is the following:

Theorem: Let (P, G, E, m, Δ) be a t-local dynamical automaton with input, $\mathcal{M}$ the free monoid over $\{(Q_i, R_i)\}$ and assume 1) that $\mu(\eta) \geq t$ for all $\eta \in \cup_{i \in I}Q_i$, and 2) that R_i is a set of dynamically equivalent states for each $i \in I$. Let $\bar{E} \subset E$. Then $\mathcal{A}$ effectively processes $\mathcal{M}$ in $\bar{E}$ iff

1. For all i, j, $\Delta(R_i, \bar{E})$ supports (Q_j, R_j)
2. For all i, j, $W_{ij} = \bar{\Delta}(R_i, \bar{E}Q_j)$ and R_i have non empty interiors.

Proof: $\Rightarrow$ Assume that $\mathcal{A}$ effectively processes $\mathcal{P}$. For any $i, j \in I$ there exists a set $S_{ij} \subset S$ such that S_{ij} supports (Q_i, R_i) and $\Delta(S_{ij}, Q_i\bar{E})$ supports (Q_j, R_j). Thus for any $\psi \in S_{ij}$, $\psi_i \in Q_i$, $\eta \in \bar{E}$ and $\psi_j \in Q_j$, there exist $R_i \in R_i$, $\sigma \in S$ and $\sigma_j \in \Sigma_j$ such that $\Delta(\psi, \psi_i\eta\psi_j) = \Delta(\psi R_i, \eta\psi_j) = \psi\Delta(R_i, \eta\psi_j) = \Delta(\psi R_i\sigma, \psi_j) = \psi R_i\sigma\sigma_j$. Thus $\Delta(R_i, \eta\psi_j) = R_i\sigma\sigma_j$.

Now $\Delta(\psi, \psi_i\eta) = \Delta(\psi R_i, \eta) = \psi\Delta(R_i, \eta) = \psi R_i\sigma$. Hence $\Delta(R_i, \eta) = R_i\sigma$.

Hence $R_i\sigma\sigma_j = \Delta(R_i, \eta\psi_j) = \Delta(\Delta(R_i, \eta), \psi_j) = \Delta(R_i\sigma, \psi_j)$. It follows that $\bar{\Delta}(R_i, \eta\psi_j) = \sigma\sigma_j$ and $\bar{\Delta}(\Delta(R_i, \eta), \psi_j) = \bar{\Delta}(R_i\sigma, \psi_j) = \sigma_j$.

52

For any $\psi' \in R_i$ we have $\Delta(\psi', \psi_j) = \Delta(\psi', \eta\psi_j) = \psi'\bar{\Delta}(\psi', \eta\psi_j) = \psi'\bar{\Delta}(R_i, \eta\psi_j) = \psi'\sigma\sigma_j$. Now $\Delta(\psi', \eta) = \psi'\bar{\Delta}(\psi', \eta) = \psi'\bar{\Delta}(R_i, \eta) = \psi'\sigma$. Thus $\Delta(\Delta(\psi', \eta), \psi_j) = \Delta(\psi'\sigma, \psi_j) = \psi'\sigma\sigma_j$. Thus $\bar{\Delta}(\Delta(\psi', \eta), \psi_j) = \sigma_j \in \Sigma_j$.

Now $\Delta(\Delta(S_{ij}, Q_i\bar{E}), Q_j) = \Delta(\Delta(S_{ij}, Q_i), \bar{E}Q_j) \subset \Delta(S_{ij}R_i, \bar{E}Q_j) \subset S_{ij}\Delta(R_i, \bar{E}Q_j) = S_{ij}R_iW_{ij}$.

Since $\Delta(S_{ij}, Q_i\bar{E})$ supports (Q_j, R_j) it follows that each set in this chain has a non empty interior.

Thus we have shown that $\Delta(R_i, \bar{E})$ supports (Q_j, R_j) for all i, j and W_{ij} and R_i have non empty interiors.

$\Leftarrow$ Assume that for all i, j, $\Delta(R_i, \bar{E})$ supports (Q_j, R_j) and that $W_{ij} = \bar{\Delta}(R_i, \bar{E}Q_j)$ and R_i have non empty interiors. Let $W_{i_n} = \bar{\Delta}(\Sigma_{i_n}, \bar{E})$, and $\bar{S}$ be the union of all sets of the form $\Sigma_{i_1}W_{i_1i_2}\cdots W_{i_{n-1}i_n}W_{i_n}$ for some sequence $i_1, \ldots, i_n$. Consider any instance $(Q_{j_1}, R_{j_1}), \ldots, (Q_{j_k}, R_{j_k})$ in $\mathcal{P}$.

First of all, note that

$$\begin{aligned}
\Sigma_{i_1}W_{i_1i_2}\cdots W_{i_{n-1}i_n}W_{i_n} &= \Sigma_{i_1}\bar{\Delta}(\Sigma_{i_1}, \bar{E}Q_{i_2})\cdots W_{i_n} \\
&= \Delta(\Sigma_{i_1}, \bar{E}Q_{i_2})W_{i_2i_3}\cdots W_{i_n} \\
&= \Delta(\Sigma_{i_1}, \bar{E}Q_{i_2}\bar{E}Q_{i_3})\cdots W_{i_n} \\
&= \vdots \\
&= \Delta(\Sigma_{i_1}, \bar{E}Q_{i_1}\bar{E}Q_{i_2}\cdots Q_{i_n}\bar{E})
\end{aligned}$$

and likewise that

$$\Sigma_{i_1}W_{i_1i_2}\cdots W_{i_{n-1}i_n} = \Delta(\Sigma_{i_1}, \bar{E}Q_{i_1}\cdots \bar{E}Q_{i_n})$$

Now consider

$$\begin{aligned}
\Delta(\Sigma_{i_1}W_{i_1i_2}\cdots W_{i_{n-1}i_n}W_{i_n}, Q_{j_1}) &= \Delta(\Delta(\Sigma_{i_1}, \bar{E}Q_{i_1}\bar{E}Q_{i_2}\cdots Q_{i_n}\bar{E}), Q_{j_1}) \\
&= \Delta(\Sigma_{i_1}, \bar{E}Q_{i_1}\bar{E}Q_{i_2}\cdots Q_{i_n}\bar{E})Q_{j_1}) \\
&= \Sigma_{i_1}W_{i_1i_2}\cdots W_{i_{n-1}i_n}W_{i_nj_1}
\end{aligned}$$

Let $\psi \in \Sigma_{i_1}W_{i_1i_2}\cdots W_{i_{n-1}i_n}W_{i_n}$. Then there exist $\sigma, \sigma' \in S$ and $\sigma_{i_n} \in \Sigma_{i_n}$, $W_{i_n} = \bar{\Delta}(\Sigma_{i_n}, \bar{E})$ and $\sigma_{i_n} \in \Sigma_{i_n}$ such that $\psi = \sigma\sigma_{i_n}\sigma'$. Thus $\sigma_{i_n}\sigma' \in \sigma_{i_n}\bar{\Delta}(\Sigma_{i_n}, \bar{E}) \subset \Delta(\Sigma_{i_n}, \bar{E})$. Thus $\Delta(\psi, Q_{j_1}) = \Delta(\sigma\sigma_{i_n}\sigma', Q_{j_1}) = \sigma\Delta(\sigma_{i_n}\sigma', Q_{j_1}) \subset \sigma_{i_n}\sigma'\Sigma_{j_1}$ by assumption. Thus $\bar{\Delta}(\psi, Q_{j_1}) \subset \Sigma_{j_1}$. Now

$$\begin{aligned}
\Delta(\Sigma_{i_1}W_{i_1i_2}\cdots W_{i_{n-1}i_n}W_{i_n}, Q_{j_1}) &= \Delta(\Sigma_{i_1}, \bar{E}Q_{i_1}\bar{E}Q_{i_2}\cdots Q_{i_n}\bar{E}), Q_{j_1}) \\
&= \Delta(\Sigma_{i_1}, \bar{E}Q_{i_1}\bar{E}Q_{i_2}\cdots Q_{i_n}\bar{E}Q_{j_1}) \\
&= \Sigma_{i_1}W_{i_1i_2}\cdots W_{i_{n-1}i_n}W_{i_nj_1}
\end{aligned}$$

By assumption, this has a non empty interior.

Thus $\Sigma_{i_1} W_{i_1 i_2} \cdots W_{i_{n-1} i_n} W_{i_n}$ supports (Q_{j_1}, R_{j_1}).

The argument can be repeated inductively to prove that $\Delta(\Sigma_{i_1} W_{i_1 i_2} \cdots W_{i_{n-1} i_n} W_{i_n}, Q_{j_1} \bar{E} \cdots Q_{j_h} \bar{E})$ supports $(Q_{j_{h+1}}, R_{j_{h+1}})$. This proves that $\mathcal{A}$ effectively processes $\mathcal{P}$ in $\bar{E}$ using support in $\bar{S}$.

The above proof has also shown the following:

The Rewriting Lemma: Let $\mathcal{A} = (S, E, \Delta, T, f)$ denote a t- local complex system with output, $\mathcal{M}$ the free monoid on $\{(Q_i, R_i)\}$, $\bar{E} \subset E$, and assume 1) that $\mu(\eta) \geq t$ for all $\eta \in \cup_{i \in I} Q_i$, and 2) that R_i is a set of dynamically equivalent states for each $i \in I$. Then for any set $\Sigma_{i_1} W_{i_1 i_2} \cdots W_{i_{n-1} i_n}$ and for any sequence $j_1, \ldots, j_k$,

$$\Delta(\Sigma_{i_1} W_{i_1 i_2} \cdots W_{i_{n-1} i_n}, Q_{j_1} \bar{E} \cdots \bar{E} Q_{j_k}) =$$

$$\Sigma_{i_1} W_{i_1 i_2} \cdots W_{i_{n-1} i_n} W_{i_n j_1} \cdots W_{j_{k-1} j_k} W_{j_k}.$$

This result demonstrates that under such conditions, the dynamical automaton functions as a linguistic rewriting system, transforming environmental transients into system transients. Thus we have an example in which the dynamics of a system gives rise to intrinsic linguistic activity. The next question is whether systems with such properties exist.

Transient induced global response synchronization, or TIGoRS, refers to the clustering, within a small, localized region of pattern space, of the responses of a complex system to a transient stimulus. TIGoRS has been detected now in a wide range of complex systems models: cellular automata[14], coupled map lattices, tempered neural networks[13], cocktail party automata.[16] The cocktail party automaton provides a particularly striking example of TIGoRS. Cocktail party automata are inhomogeneous, asynchronous, stochastic, adaptive cellular automata. Inputs to such automata originate within the pattern space, but are sampled stochastically at varying rates.

Table 1 displays the results of four trials, each consisting of 100 presentations of a stimulus to a single inhomogeneous, asynchronous, adaptive cocktail party automaton using recognition mode. The stimulus varied between the trials. Displayed are the mean (variance) of the Hamming distances between responses and a fixed response template (S), between responses and input pattern (P), between each individual pair of responses (DP) and between each individual pair of randomly generated surrogate patterns having an identical distribution of pattern norms to that of the original sample (**RP**).

Ham Dist	TIGoRS			
	Strong	Weak	Pseudo	Absent
S	4 (1.4)	31 (3.7)	18 (2.3)	48 (0.9)
P	3 (1.2)	28 (2.0)	21 (4.2)	44 (4.7)
DP	4 (2.1)	31 (5.5)	16 (3.9)	47 (5.1)
RP	39 (6.0)	49 (5.0)	17 (4.4)	49 (4.9)

Table 1: A Sampling of TIGoRS

One observes the localization in pattern space of the output patterns which provides direct experimental evidence of this linguistic behavior. TIGoRS provides a weaker form of linkage between environmental and systemic transients than that described above. The patterns in the response set need not be dynamically equivalent. Nevertheless we still achieve a similar kind of linguistic structure.

7 Conclusion

I began the main body of this paper with two questions:

1. What is the relationship between dynamical systems theory and natural language?, and
2. Which dynamical properties are necessary and sufficient to give rise to linguistic behaviour and structure in the observable behaviour of a system?

An answer to the first question does not yet exist in the literature. Nevertheless, restricting it to the relationship between dynamical systems theory and formal language theory, one begins to see that there are very deep interconnections present between these two fields. As revealed in Moore's work on generalized shifts, deep problems exist concerning the epistemology of dynamical systems, in particular, serious questions are raised concerning our capacity to obtain global knowledge about the specific behaviors of classes of such systems. It is just these kinds of questions which psychology and psychoanalysis attempt to answer. We are as yet a long ways from determining the class of dynamical system within which to situate psychological processes. Nevertheless it is highly likely that they would lie at least within that containing the generalized shifts, or more general systems. Thus we must be cognizant of the possibility that our ability to perform long term predictions of behavior may face serious limitations. This limitation of predictability cannot be overcome by the use of a natural language on account of the Church-Turing thesis.

It should also be clear that the relationship between these formal languages and natural languages is far from straightforward, and need not preserve dynamical structure at all. As a result one needs to at least exercise caution when attempting to utilize natural language in a rigorous manner to describe and to understand dynamical phenomena.

Crutchfield's work demonstrates that some dynamical phenomena carry a corresponding linguistic counterpart, reflected in particular by an increase in linguistic complexity across a phase transition. In a fundamental way, the language required to adequately describe a phenomenon in a function of the phenomenon, and careful attention to this can illuminate subtle differences which might otherwise pass unnoticed.

The approaches above all involve the use of an externally derived language to symbolically represent phenomena occurring within the system in question. Thus they do not truly address the second question which seeks intrinsic linguistic structure in a dynamical system. Dynamical automata provide the beginning of an answer to this question. Here it can be demonstrated that certain classes of dynamical systems behave in such a way that the coupling of environmental stimulus and system response is akin to that of a system which reads the envirnomental stimulus as a letter in a language and replaces that letter by a new letter, expressed as the system response. In such a system we have an intrinsic linguistic structure. One can conjecture that such an intrinsic linguistic structure may be a prerequisite for the eventual emergence of an explicit linguistic structure in the form of a natural language. Of course such a conjecture is rank speculation at present, yet presumably our perception of world through language is effective because the world itself possesses behavior which lends itself faithfully and homologously to a linguistic description. If so, then an understanding of the necessary and sufficient conditions which underlie the generation of such linguistic behavior by a dynamical system will help us to restrict the classes of dynamical systems to be studied as models of psychological systems.

In addition, as has been argued elsewhere[19], such a dynamic linguistic structure obviates the need for explicit representations of knowledge by a complex system, and thus provide a vehicle for subsymbolic processing of the kind thought important in the behaviour of collective intelligences. One can conjecture that such subsymbolic linguistic structure may also underlie other forms of adaptive behavior, such as that which occurs at lower sensorimotor levels, the study of which is proving important in the design of autonomous agents such as robots.

References

1. F.D.Abraham and A.R. Gilgen, eds., *Chaos Theory in Psychology* (Praeger, Westport,CT, 1995).
2. J.J.Collins and I.N.Stewart, *Nonlinear Science* **3(3)**, 349 (1993).
3. J.Crutchfield and K.Young in *Complexity, entropy, and the physics of information*, SFI Studies in the Sciences of complexity, Vol. 8, ed. W.H.Zurek (Addison-Wesley, Menlo Park, CA, 1990).
4. R.Devaney, *An Introduction to Chaotic Dynamical Systems* (Addison-Wesley,Menlo Park, CA, 1989).
5. I. Golani, *Behavioral and Brain Sciences* **15(2)**, 249 (1992).
6. S. Guastello, *Chaos, catastrophe, and human affairs* (Lawrence Erlbaum, Mahwah, NJ, 1995).
7. N.Kopell and B.Ermentrout, (1988) *Mathematical Biosciences* **90**, 87 (1988).
8. C.Moore, *Nonlinearity* **4(2)**, 199 (1991).
9. M.P.Paulus and M.A.Geyer, preprint (1993).
10. K.Petersen, *Ergodic Theory* (Cambridge University Press, Cambridge,1993).
11. R.Robertson and A.Combs, *Chaos Theory in Psychology and the Life Sciences* (Lawrence Erlbaum Associates, mahwah, NJ, 1995).
12. W.Sulis in *Proceedings of the International Joint Conference on Neural Networks '92* Vol. III, 421 (IEEE Press, Baltimore, 1992).
13. W.Sulis in *Proceedings of the World Congress on Neural Networks '93* Vol. IV, 448 (Lawrence Erlbaum Associates, Mahwah, NJ, 1993).
14. W.Sulis, in *1993 Lectures in Complex Systems*, Santa Fe Institute, eds. D.Stein and L.Nadel (Addison-Wesley, New York, 1995).
15. W.Sulis, *World Futures* **39**, 225 (1994).

16. W. Sulis in *Advances in Artifical Life, Lectures Notes in Artificial Intelligence 929* eds. F. Moran et. al. (Springer-Verlag, New York, 1995).

17. W.Sulis and A.Combs, eds., *Nonlinear Dynamics in Human Behavior* (World Scientific, Singapore,1996).

18. W.Sulis, W. Poster presented at Alife 5, Kyoto, Japan. May, 1996.

19. W.Sulis in *Nonlinear Dynamics in Human Behavior*, W.Sulis and A. Combs, eds. (World Scientific, Singapore, 1996).

NON LINEAR DYNAMICS IN LANGUAGE
AND PSYCHOBIOLOGICAL INTERACTIONS

FRANCO ORSUCCI

Department of Psychology and Institute for Complexity Studies,
Rome International University,
Viale dell'Università, 27 - Rome, Italy, I-00185 - E-mail: OF@iol.it

Language and thinking give us access to what is usually called *natural and social reality*. Language and thinking can be considered as parts of a *semiotic universe* of different entities from which emerge different subsets. Some of these can have peculiar functions in interpersonal interactions and biological transductions. Nonlinear studies at the *morphological level* of language are opening new perpectives in this area of the Mind-Sciences.

1 The Psyche-Soma Paradox

Since Samuel Taylor Coleridge coined the term 'psychosomatic' in 1796, the history of this area of studies on "the mysterious leap from the mind to the body" has been developing in balance between several paradoxes. It has been presented as a holistic discipline, interested in the study of the mind-body global system, but it has often been reduced to the deterministic search for hypothetical events in the virtual psyche-soma interface.

Sigmund Freud gave us a framework to manage this paradox: in the "*G-Draft, Melancholy*", a part of his letters to Fliess[1], he defines a model which is a sort of 'topological phase-space' connecting the Object to the external world, the inner world, and the biological body .

The elements of the model are allocated in a space divided by the intersection of two boundaries (the Ego border, and the psychosomatic border): 1) an object, in the external world; 2) an object in a favorable position, outside the Ego, in the body; 3) an end-organ, a somatic source, and a spinal center, in the body-Ego; and 4) a psychic sexual group, in the psychic Ego. These elements are the main stations in a circuit performed by a vector: sexuality, with its drive, goal, source and eventual obstacles.

60

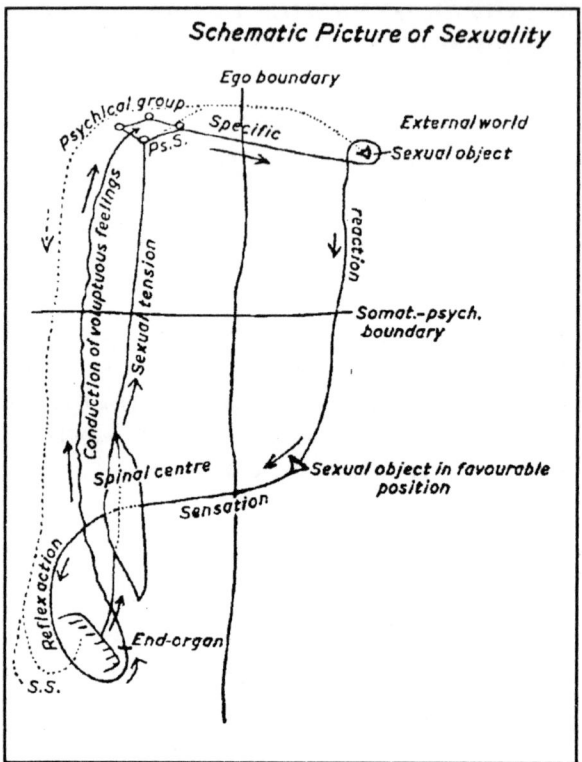

Figure 1: G-Draft Diagram (from Freud, 1895).

It is amazing how this dynamical and complex model has been neglected in the history of psychoanalysis and psychosomatics. This model is entirely based on drive, movement and loops, while other famous Freudian models are more deterministic and static.

Freud was well aware of the features of psychic dynamics for which he coined the specific term *'Überdeterminierung'*, or over-determination: every unconscious formation is a multicausal derivative; and every derivative is linked to multiple unconscious elements, which can be organized in several sequences of meaning, each with its own coherence. The *'Überdeutung'*, the over-interpretation, is the logical consequence: in this non-euclidean and interwoven stratifications of meaning, several interpretations are allowed and there cannot be simple defined criteria to determine the correct one.

In the same years the American linguist Charles Sanders Peirce[2] was exploring the same features of the semiotic universe asserting that meaning is an *'interpretant'* which acts like an interpreter who asserts that a stranger means the same thing that he is saying, but in another language. Every interpretant has its own interpretants, and this yields an infinite chain of triadic links of translations (symbol-interpretant-object), an endless fugue in the semiotic universe.

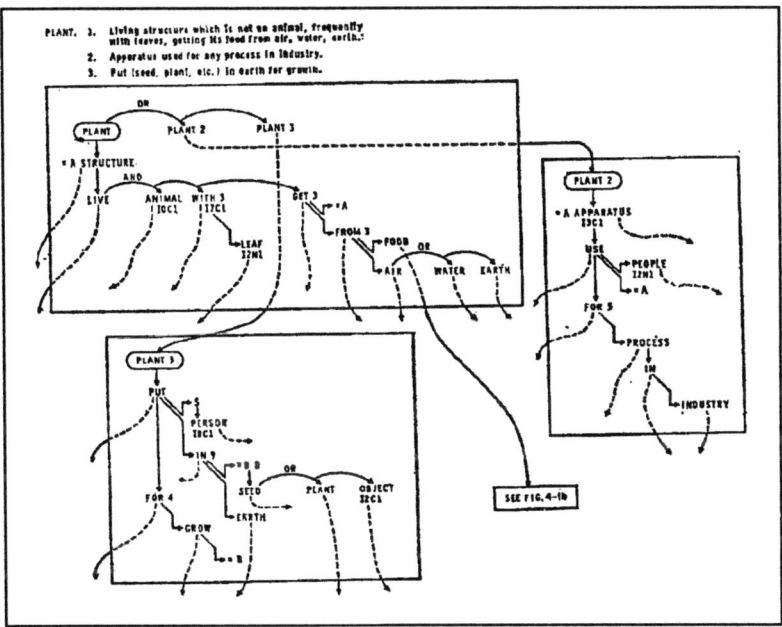

Figure 2: A small part of the semiotic universe, a semiotic map of the term *plant* (from Quillian MR,1968).

Nonetheless, the first steps of the "classical" psychosomatics scientists[3,4] can be recognized as attempts to define linear deterministic schemes that explain somatic

62

diseases as caused by psychic dynamics or structures. The foundations of the "classical era" of psychosomatics are based on the assertion that a specific conflict or a specific personality structure could produce a specific bodily disease. The term "psychosomatic" was virtually equated with *"psychogenetic"*, and this assumption was applied in the study of a small group of chronic diseases of uncertain ethiology (peptic ulcer, bronchial asthma, essential hypertension, thyrotoxicosis, ulcerative colitis, rheumatoid arthritis and neurodermatitis), the so-called "holy seven".

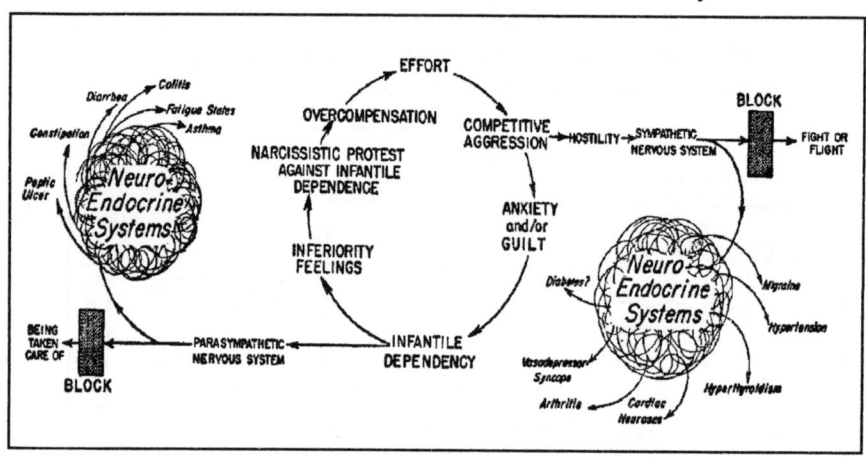

Figure 3: Psychosomatic interactions (from Alexander F, 1950).

These theories were a derivation of hierarchical determinism according to which the psyche (at its higher levels) rules over the body. It was also an attempt to apply the Freudian model of the 'transference neuroses' (mainly hysteria[5]) directly to psychosomatic syndromes. In this way the 'actual neuroses model', the biological complexity of the Id[6] and the circular interactions mind-body-environment, as outlined in the "G-Draft", were all left behind.

However, the progress of science regarding the various domains in the body-mind framework opened the way to recognizing that the "psychosomatic field" can be understood only in terms of multicausal processes and feed-back loops. Modern psychosomatics became aware that we can only understand the psychosomatic processes from a complex *"biopsychosocial"* perspective[7]. The earlier notion of psychogenesis was discarded, as linear models of psychosomatic causality were gradually replaced by cyclical and interactional models that suggested somatic-psycho-somatic sequences[8].

We cannot simply determine the ultimate unique origin of a disease because once a disease is triggered or activated, the individual is changed - physiologically and psychologically. The "inseparability and interdependence of psychosocial and biological (physiologic, somatic) aspects of humankind"[9] are a definitive basis for today's psychosomatics. In this complex framework every disease can be regarded as "psychosomatic" in its genesis and evolution.

2 A Single Case Study

A single case-history, in its endeavor to present a life and the repercussions of a disease, may support a renewed sense that complex neural and psychic functions (and their disorders) require detailed and non-reductive narratives for their explication and understanding. One case generally does not support true scientific findings, but it can be used as a start in indicating new directions in research, just for its capacity in mirroring complexities.

Single case studies, since the time of Groddeck[10], were the basic tools to approach clinical and theoretical formulations in psychosomatics. But in the last thirty years they were discarded as non scientific, in comparison with studies on large populations. It has only been in the past two or three years that single case studies were recognized again as worthy of publication in the main international journals of this field. This re-evaluation of clinical stories was caused by the acknowledgment of the complexity and the partial uniqueness of the psyche-soma-environment interactions in each human individual.

The clinical history that we are going to exam, just for its atypical features, can be a source of interrogatives, curiosities and new thinking. It is a single case of *Behcet's Disease* that has been studied through several consultations during a period of 15 years. The evolution of the disease is related to the subject's important life-events. The cognitive patterns and the psychopathological features of the subject were thoroughly studied and his couple and family relationships were related to the somatic disease and cognitive patterns[11].

Behcet's Disease (from the name of a Turkish physician) is characterized by three primary components: iridocyclitis (historically with hypopyon), aphthous lesions and ulceration of the mucous tissues[12]. Erythema nodosum arthropathy and thrombophlebitis often accompany these manifestations. Central nervous system involvement, most often due to necrotizing vasculitis, may be the most protean manifestation of the disease, leading to death. The underlying disease mechanism in all organ systems is an *occlusive vasculitis.*

The disease in some geographic areas (Japan, Near East) is observed to have rather high prevalence rates. The present view of aetiopathogenesis focuses on the abnormal increase of circulating immune complexes (mainly IgG complexes) against mucosal and/or microbial antigens, which accumulate in both small and larger blood vessels, eliciting a severe vasculitis with necrotizing tissue damage[13]. Numerous immunologic abnormalities may be detected, including the presence of autoantibodies against affected tissues and of circulating immune complexes.

The patient was observed in several consultations during a period of 15 years, including five years of regular consultation and ten years of follow-up. He works as a scientific researcher and, at the time of the first consultation, was in his early 30s. He was single and came from a middle class family, where he was the third son of four. His family and personal life seemed normal until the suicide of his father, who was deeply depressed and debilitated by a global ateriosclerosys.

At the time of the first consultations the patient had a very "fast" private life, spending every night in clubs and bars, drinking, dancing and having many relationships. His father's death caused an escalation of his fast paced life style. His cognitive patterns and mood were almost always tuned to a *manic* level.

Six months after the death of his father he started to present episodes of relapsing purpura on his feet and legs. He also noticed recurrent attacks of aphtous ulcers.

When these symptoms began he refused to have medical consultations, but the patient was, in some way, conscious of his profound distress and he asked for his first consultation.

He presented a quick *way of speaking* marked by frequent jokes, which gave the impression of rich and creative thinking. But there was also an underlying *state of mind* of emptiness and desperation. The patient refused to go to a hospital and consult a well known clinician, and he increased his frenetic way of life; however, he continued psychotherapeutic consultations.

It was only after several stronger episodes of purpura and aphtas, several months after the first consultation, that the patient decided to start a long clinical screening. At the end of the clinical trials and the immunological tests, the diagnosis was defined as 'Behcet's Disease'. Laboratory results were characterized by the presence of immunocomplexes, slight renal damage, and autoantibodies against endotelium and DNA. He was also recognized as an HLA-B5 type, having a group of tissue antigens that is frequently related to Behcet's Disease.

The first therapy he underwent was based on high doses of corticosteroids, but the results were scarcely significant. He continued his "night life" although conscious that it was a kind of slow suicide. The pharmacological therapy was then changed using first colchicine, and then levamisole, but the results were only slightly better. During his sporadic consultations with me the patient became more evidently *depressed*. However his temporary insights were not stable enough to enhance his motivation towards a regular psychological or psychopharmacological treatment. The path towards his death seemed inevitable.

In these days he spoke quite a bit about a reproduction of a famous painting by Goya: *Saturn*. He said that he used to observe this image on a wall of his bedroom, and it was not clear to me if this was also the content of some nightmares, a mental imagery, or merely a real poster on the wall.

This scary and bloody painting of the giant Saturn devouring his sons was painted in 1823, and is regarded to be a representation of *"Time, the Destroyer"*, but also of melancholy and senescence. Another critical interpretation of this painting as part of the so-called *"Goya's Black Paintings"* sees in the devoured body a woman, and the cannibal god as representing the sexual appetites[14].

In those days he said also that the patches on his legs, produced by the purpura crises, reminded him of similar patches on the legs of his father, produced by the ateriosclerosys. A trait functioning as bridge towards a pathological identification.

Two years after the beginning of treatment, he met a young French woman who was an "au pair" for his sister's family. The woman had worked as a diplomated psychomotricist with psychotic children in France. She was in a period of "burnout" due to her work, and also of distress produced by some problems in her family life. For these reasons she decided to take a sabbatical in Italy.

They began a relationship and she was eventually informed of his physical and psychological conditions. They had some joint consultations with me, and she expressed concern about his disease and the danger to his life. At this point she acted as a "psychological conscience" for him, always reminding him of the psychological bonds between his present situation and his previous history and life events, the death of his father.

With her help the patient *regulated* his life-style, and this produced a slight amelioration. He began a new cycle of medical consultations in France. The initial diagnosis was confirmed, and the original treatment was supported by a diet which excluded milk and all its derivatives. Three years into treatment, they were married in beautiful ceremony in her home village in France.

During a consultation shortly into their married life, they related a peculiar aspect of their *patterns of thinking during the night*. Before marriage she had had a rich dream life which was completely interrupted, after the marriage, by "blank sleeping". The patient, on the contrary, recounted that while before the marriage he couldn't remember any dreams, afterwards he "started to dream almost every night". At first he related complex and scary nightmares but, after about six months, more common dreams.

His wife strongly wanted a child and she succeeded in having a lovely baby boy after a year of marriage. During a consultation with me he stressed that "being a father" was very important to him. This was concurrent with another step in his amelioration. All of the ulcers disappeared, and later he had only brief episodes of purpura, often as a result of the stress caused by his business that took him to several countries of Europe.

Today, almost fifteen years after the outbreak of the Behcet's Disease, the patient still lives and works. His last purpura episode occurred two years ago. He has two children now and still lives with his French wife. She is working again as a psychomotricist, but no longer with psychotic but with handicapped children. One and a half years after their marriage (about four and a half years after the beginning of treatment), she asked for some consultations with me for several episodes of depression, alternating with periods of low-back pain.

The relapsing cycle of the patient's Behcet's Disease has had interesting coincidences with important life events[15], important changes in his attachment bonds[16, 17], modifications in his patterns of thinking[18, 19, 20], and in the psychic conditions of his partner.

There is also an evidence for an over-determination in the material of this clinical history. There are some biological events, and measures of biological variables; there are some antecedents in the family of the patient which could contain both biological and psychological value; there are some important stressful life events; there is some change in the subject's and couple's cognitive assets; and there is also what seems like a sort of transference of psychic functions.

It was not possible, with this patient, to work in the traditional psychoanalytical setting, but it was necessary to keep in mind a psychoanalytical function. So, it was never interpreted, for instance, the conflicts with his father, and the revenge of the dead-father, Saturn. He was aware enough to offer these interpretations, maybe to disqualify them. We may say that the psychoanalyst concretely acted as *"helper"* and after his meeting with the French woman, acted as a *third-witness* (also in a Peirce-Lacan sense) in a complex process. Then she became "the helper" as a symbiotic therapist, by giving him reparation through a transference of functions. In this way, a triadic configuration was formed , and concretely built some basic conditions for the mental representation of suffering and pain.

This leads to a basic feature of the psychosomatic mind: mental operations are performed on *real* objects (not on their psychic derivatives) by an operative thinking, 'pensée operatoire'. This implies that any psychotherapeutic process in the psychosomatic field has to move, in complex loops, between real, imaginary and symbolic. The conditions for access to the social expression of suffering, and its substituting to somatic suffering, should be built into reality.

This clinical history arises some questions: How many variables were involved in this process, how many degrees of freedom ? Were they organized in a few discernible patterns ? Is there a possible hierarchy among them ?

3 The Nachträglich Effect

This is a term frequently used, and also underlined by Freud himself, since the 'Project[21], translated into English as 'deferred action' and in French, as used by Lacan[22], it means 'après-coup'. It is related to the assumption that time and cause in the psyche do not have linear evolutions: the present acts on the past and the past on the present. And this also brings about the important consequence that every time one recalls a memory it is not the same. Or that, in the case of traumatic scenes, is the second trauma which gives to the first a part of its pathogenetic power.

Related to this conceptual area there is also the Jung's definition of 'Zurückphantasieren', meaning deferred fantasies, a way to build reinterpretations of the past by using imagery. This assumption gives more radical implications to the *'Nachträglich effect'*, because it suggests that by using fantasy we can build memories.

In this way we are introduced to the complexities of the *psychic work* ('psychische Verarbeitung'), so well described by Freud[23] in the C Appendix of "Inhibition, Symptom, Anxiety", as "the endless psychic efforts to link bodily excitations to mental representations".

This kind of approach is evident also in modern psychoanalysis as in Bion's "model" of "$\beta <-> \alpha$ *oscillation*"[24, 25] (also in the variation "$PS <-> D$"). In this "model" is represented the continuous oscillation between raw sensory data and reprentations, as precursors of thinking. This "model" often originates a deterministic pressure of transforming $\beta \rightarrow \alpha$, but sometimes it comes with the acknowledgment that the oscillation $\beta \leftrightarrow \alpha$ respects a natural balance.

As stated by Peirce[26] the elements of language (and so also of thinking) could be divided into three main categories: *indexes, icons and symbols*. Lacan, used this triadic distinction to define three domains of our psychic life: the Real, the Imaginary, and the Symbolic. He added that our reality is built by all the three, together, as in the topological structure called the *"Borromean knot"*: if you cut one part of the knot all of the others fall away as well. It does not make sense to ask where the beginning of this knot is: it could be only a more cultured version of "the chicken and the egg" problem.

Walter Freeman asserts [27,28] that from a psychobiological perspective he can distinguish two forms of self: the *"objective self"*, constituted by brain dynamics, and the *"subjective self"*, based on awareness. These two forms fuse in the process of perception "by which global patterns of activity come briefly into focus at a point in a trajectory, collecting all the residues of past experience, and influencing by reactive transmissions all that is to come in the life span of an individual".

This does not imply the inexistence of mental representations, but it does need a reformulation of the concept. Mental derivatives (à la Freud) should not be seen as "ghosts in the machine", 'homunculi' with their sites on the cortex. Mental representations should be considered as *dynamic structures*, as Fairbairn suggested years ago in the psychoanalytical field[29]. This theory also provides a basis for explaining the development of symbols as stable attractors in the dynamics of the psyche-soma continuum. Chaotic fluctuations allow the presence of random processes which permit the flexibility of biological systems, to escape local minima in constrained optimizations. They also enable a very efficient form of data compression in memory, in which the complexity of a given CNS state is reduced to the topological form of key attractors.

4 Pensèe Operatoire

One of the most salient features of the subject examined in the clinical history is his disorder in the expression of emotions, connected with a lack of mindedness and preference for attention to the external world. These traits were evident mainly at the beginning of his clinical history. They are collected in a specific personality construct which is called *"alexithymia"* (meaning literally 'no word for emotion'), and has an important role in modern psychosomatics[30].

The term alexithymia was first coined by Nemiah and Sifneos to define some features noted in the so-called "psychosomatic subjects", but the term's definition soon expanded to a personality construct including a variety of other conditions. We can define the main alexithymic traits as:

1) difficulty in feeling, recognizing and finding appropriate words to describe feelings;

2) poor fantasy life and elaborate description of trivial environmental details;

3) poor dream life;

4) interpersonal relationships in balance between strong dependency or avoidance;

5) a life-style centered on action;

6) the interviewer finds the subject boring, dull or void.

This personality construct is now a well defined heuristic instrument and has a specific measure in the T.A.S., a well validated rating scale developed by Taylor and his group[31]. It is very interesting to note that the prevalence of alexithymic traits has been found in several other conditions, such as: drug addiction, somatoform disorders, postraumatic stress disorders[32], chronic pain[33], obesity and anorexia[34], major affective disorders and masked depression[35], sexual perversions[36], sport addiction, degenerative diseases and long hospitalizations[37].

The idea that a kind of variation in brain organization is linked to alexithymia is supported by the demonstration[38] of higher scoring in tests among commissurotomy patients than in control groups. It is demonstrated as a deficit in the interhemispheric transfer in alexithymic men with postraumatic stress disorder[39], and also in children and adults with evidence of lesions in the right hemisphere[40]. Also an association between alexithymia and left cerebral lateralization has been found in a sample of young adults with intact brains[41].

Antecedents of the alexithymia construct are the works of Marty and de M'Uzan[42] who proposed the concept of *"pensée operatoire"* to define the lack of imagination and the externally oriented thinking of the psychosomatic patients which they observed.

This definition of "pensée operatoire" is obviously, if not explicitly, derived from the works of Jean Piaget[43]. He was the first to define this cognitive asset, with its specific logical features and the first to propose a specific datation in the human development. Following his observations these are the main features of "operative thinking":

1) there are complex cognitive structures allowing sophisticated logical operations,

but there is a lack of abstraction from physical objects (Piaget lists nine kinds of logical operations available at this stage);

2) the capacity to understand the difference between experienced events and possible events is reduced to real experiences;

3) the formal logic and the capacity to define conceptual hierarchies are not used;

4) there is a lack of capacity to assume a meta-position or to combine several vantage points;

5) the difficulty in reaching abstraction, and the interruption of the bonds between what is real and what is possible make the use of irony, play and creative thinking peculiar or difficult .

The normal development of this stage in the maturation of thinking is posited by Piaget to be in the first adolescence[44]. Some authors tried to define an integration between Piaget's studies and the psychoanalytical studies on the expression of emotions[45]. They conceptualized a sequence of stages from a simple awareness of undifferentiated bodily sensations to an awareness of complex blends of feelings and the capacity to appreciate the emotional experience of others. There is clinical evidence that shows that people who could be placed (by regression or arrest in development) in the lower stages of this model have a rigid asset in the regulation of emotional and psychophysiological balances. If these subjects are going trough a stressful time they are prone to regulate their states manipulating people, consuming alcohol, drugs and food. They also tend to experience some "strange physical distress", caused by an instability of the autonomous nervous system, and undergo a somatic disease or a casualty.

The problem of psychophysiological regulation has precursors in the Freudian formulations of the 'Project' on the *"principle of constance"* and the *"principle of neural inertia"*[46].

In his attempt to extend the physical principles to psychology, Freud followed the works of Fechner[47]. He proposed a regulatory rule for the secondary (or conscious) processes (maintaining a constant level of excitement); and another regulatory rule for the primary (or unconscious) processes (the tendency towards a complete discharge of energy). In his later works these principles were sometimes unified in the *Nirvana Principle* [48], with various theoretical and clinical problems. It was clear, anyway, that no living system could survive following the neural inertia or the Nirvana Principle.

Cannon[49] introduced the term *homeostasis* to describe the way in which the personal environment is maintained in a relatively steady state by means of numerous control systems that regulate the functioning of the organs and tissues of the body.

According to this theory the fluctuations of bodily variables are due primarily to external influences which perturb the basic steady state. This homeostasis principle is partially similar to the Freudian constancy principle.

In recent years, several psychosomatic investigators[50,51,52,53] have begun to use the general systems theory and the cybernetic principles of feedback, self-regulation, and disregulation to conceptualize the psychobiological mechanisms underlying health and disease. In this way they expanded the concept of Cannon's homeostasis, without criticizing or renouncing his theory.

Now we know that if the theory of homeostasis relating health to constancy supports the idea that disease and other perturbations are likely to cause a lack of regularity, Non Linear Dynamics predict just the opposite: that a variety of diseases or disturbances lead to a loss of complexity.

The alexithymia construct could be better understood as a situation in which a loss of complexity has been produced, leading to a rigid and weak equilibrium, which could be easily disrupted as in the extreme example of sudden cardiac death[54,55,56]. If emotional expression can enhance the complexity of cognitive and physiological systems[57] a logical consequence should be a re-definition of alexithymia and related studies in the light of Non Linear Dynamics.

The Skarda-Freeman model[58] of the olfactory bulb propose chaos as the basic form of collective neural activity for all perceptual processes. Its function is threefold: as a controlled source of noise, as an access to previously learned patterns, and as one to learning new patterns. Novel stimuli increase the correlation dimension in the bulb, while familiar stimuli reduce it.

Changes in the fractal dimension have already been described in association with a diverse range of neuropsychological events (including sleep, performance of cognitive tasks, meditation, anesthesia, and dementia) and fractal analysis has been reported to enhance the computerized detection, prediction and localization of epileptic activity [59,60]. Epileptic attacks are accompanied by a loss of variability in the brain's electrical activity[61], as occurs in Parkinson's disease and old age, though in different ways. Using strategies of anti-control we may also have access to therapies directed at degenerative and chronic disorders [62,63].

5 Biology of Relations

One of the impressing features of the clinical history that was previously examined is the variation of the biological conditions of the patient concurrent with variations in his social life such as the loss of his father, his wedding, the birth of his son. This feature of the human psychobiological functioning is well known from antiquity, but there is a certain scientific study that is relatively recent.

The research conducted by Reneé Spitz on institutionalized infants and their anaclitic depression is a cornerstone study[64]. It was followed, during the fifties and sixties, by Engel and Schmale[65] who observed a higher risk of disease in individuals who were unable to cope with separations or object loss. These subjects developed a "giving up/given up complex" whose biological implications have been widely proved and explored in recent years: profound disregulations in the endocrine, immunitary and neural mediatory systems[66].

Following Bowlby's[67,68] observations on attachment behavior, psychosomaticists mainly studied the mother-infant bond as a homeostatic organization regulated mainly by its "inner" emotional signals. In a series of studies on rodents and primates, Hofer[69] and other developmental biologists demonstrated other hidden regulatory mechanisms. These hidden multiple and pre-emotional factors act on different sensorial channels: nutritional, olfactory, tactile, thermal, visual and vestibular.

For instance, the importance of bodily contact and tactile stimulation was demonstrated by the finding that a decrease in the levels of growth hormone in separated rat pups can be prevented by stroking their skins with a brush[70]. Also, a 30% reduction of heart rate following separation could be prevented by providing the pups with a feeding. Further, the body temperature of infant rats, which is determined largely by the body temperature of the mother, has been shown to regulate levels of brain peptides, nucleic acids and amines, all of which are reduced if the young rats are prematurely separated from their mother[71]. The olfactory stimuli are also involved in the regulation of crucial aspects; for instance infant rats are unable to locate the nipple in absence of a pherormone secreted from the mother's areolar glands, whose secretion is stimulated by the suckling.

There is evidence of a multiplicity of regulatory mechanisms, most of which remain *hidden* from a passive third observer, outside of the "nursing-couple", and can only be discovered in experimental contexts. Also, in humans these kinds of regulatory mechanisms are operated on different levels than through overt emotional expression and language. They, anyway, play a basic role in our growth and health. The works by Margareth Mahler[72] on the vicissitudes of human symbiosis, and the contributions of Bowlby[73,74], Winnicott[75] and Kohut[76] can be a general framework for these findings. The interactions between the self and its self-objects (also as transitional-objects or as sensation-objects) regulate the psychosomatic balance.

Synchronizations and desynchronizations of biological rhythms in individuals and groups can be triggered by a variety of interpersonal relationships and social demands[77,78]. A common experience of this kind of hidden psychobiological mechanisms in everyday life is related to the well known synchronization of menstrual cycles of women living in the same ambient as, for example, roommates in a college.

The interchanges in our everyday relationships can be considered as formed by a bulk of semiotic entities, and only a small part of them are formed by symbols, while the most are indexes (using Peirce's and Eco's terminology), and we are unaware of almost all of them. They contribute to the relational Id and the environmental Id of which we are unaware. They act on our minds and bodies.

6 Morphological Patterns in Language

We examined several clinical experiences and empirical studies which suggested the importance of subsymbolic and often "hidden" interactions. A better definition of their presence and organization could be important. A possible approach could lay in studies on sensorial and perceptual interaction, while another important approach resides in studies on subsymbolic patterns in language and communication.

We could investigate some forms in language, its *morphology*. This term was originally used in biology, but since the mid of this century has been used to describe the type of investigation which analyzes all those basic 'elements' which are used in language. This kind of 'elements' in the form of a language are technically known as *morphemes*. At this level of the semiotic universe we can study the forms built by different combinations of basic elements.

More than four decades ago Shannon[79] established the fundamental concept of entropy for human writings, which is a meaningful measure of information content. In a general sense, a novel, a poem, a play of music, a speech transcription, a computer program, a DNA or protein sequence, etc. can be regarded as a one-dimensional string of elements. Grassberger[80] has devised a scheme to estimate Shannon's entropy for very long strings. The most commonly used method to probe the correlation in a string of elements has been the Fourier power spectrum, but recently other methods have been proposed.

A new measure of complexity which is a generalization of entropy, was introduced by Y.C. Zhang[81] as the exponent α which is related to the usual power spectrum exponent β by the following relation:

(1)

$$\alpha = \frac{\beta + 1}{2}$$

where β is defined by the power spectrum of v, S_u $(f) \sim 1/f^{\ b}$. The value of $1/2\,\alpha$ implies that the string is not long range correlated, and it leads to white noise, while $\alpha = 1$ corresponds to the scale invariant $1/f$ noise, which is also called the maximum complexity limit.

Along with this kind of approach Schenkel et al[82] analyzed several long texts from H.G. Wells novels, to computer programs, to the Bible and Koran (in various translations). They found several values of complexity ranging from $\alpha \simeq 0.52$ ("The Time Machine") to $\alpha \simeq 0.986$ (a computer program), and concluded that "one may wonder what is behind the particular shape of a writing". They reached this conclusion by a preliminary transformation of letters in binary numbers which can generate a "random walk model".

M. Amit et al[83], with an analogous approach, studied the long range correlations in various translations of the Bible (as a sample text) by mapping them onto a one dimensional random walk model. They concluded that the exponent α varies from one language to another, while it takes its maximal value in the original version i.e. Hebrew for the Bible: this means that translations reduce complexity. They assert also that correlation may be due to two reasons: correlation between ideas, and long range correlation due to syntax structure. They suggest to change name to the a value which could be better defined as "rigidity", as the more the syntactic structure is rigid (as in computer programs) the higher are a values. They, finally, suggest that hidden coded (encrypted) information can be embedded in texts as has been shown in studies on equidistant letters sequencing (ELS) on the Hebrew Bible.

A phase-space approach has been applied[84] to the analysis of transcriptions from speech of psychiatric patients. This analysis implied a preliminary translation of letters in numbers and, after the construction of the phase space plot, measurements of the Hausdorff dimension and Lyapunov exponents. This method highlighted some interesting differences between pathologies: dimensions for Depressed patients where evidently different from Schizophrenics and normal controls.

Similar time-series have been also analyzed[85] by implementation of another method: the Recurrence Plot Analysis (RPA).

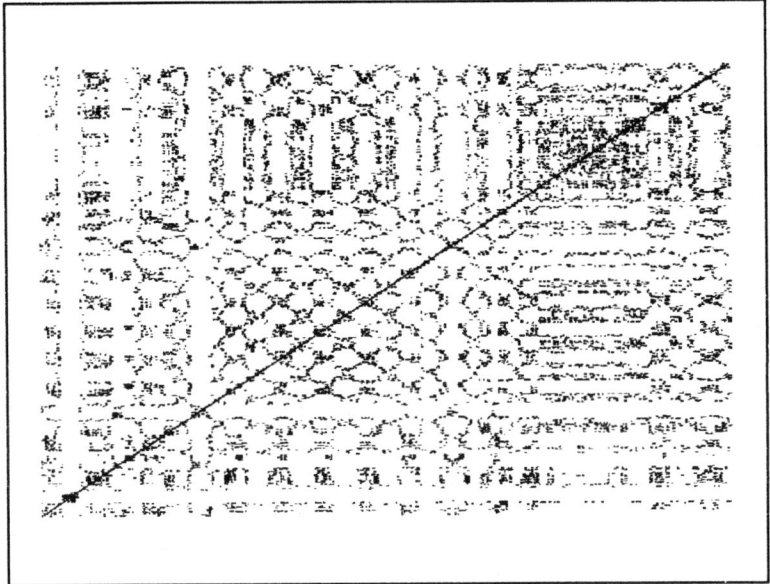

Figure 4: Recurrence Plot of a speech transcription (Orsucci & Giuliani, 1997).

This method was first introduced in the physics by Eckman, Kamphorst and Ruelle[86] and widely used in biomedicine by Webber and Zbilut[87]. Reliable metrics are needed, in order to evaluate the presence of subsymbolic patterns and compare the different texts originated by patients' speeches.

These approaches are very promising to study morphological patterns in language. The general framework of these latter studies is originated by the idea that each subject has a personal blueprint in speech which, in semiotics, is called 'idiolect'. The idiolect represents the general structure of the individual's linguistic system. Human relationships could be defined as idiolects' interactions and couplings, and they happen on multiple levels: symbolic and subsymbolic (icons, indexes-information). In this direction the work by Nicolis & Katsikas[88] on chaotic dynamics of linguistic-like processes at the syntactical and semantic levels is an attempt to model these linguistic complexities, in the pursuit of a multifractal attractor.

The morphological level of language and thinking can be posited at the interface with the neurobiological structures. This is the level in which the bidirectional transduction processes of information between mind and body are always active. Couplings of mophological-informational attractors can drive the body-mind dynamics, as well as been implied on some levels of the interpersonal and environmental interactions.

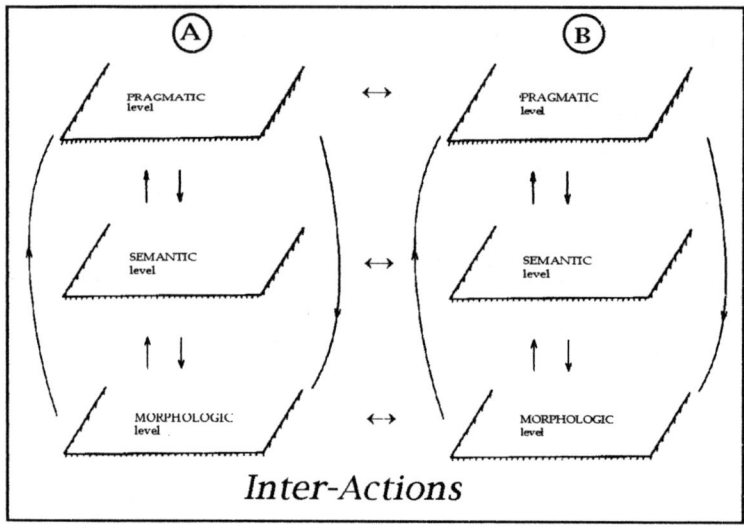

Figure 5: diagram on multilevel, interpersonal, semiotic interactions . Subject A and subject B are represented by multilevel linguistic universes in multiple couplings (Orsucci, F. 1997).

7 Dimensions

We have examined how some dimensions of language can be implied in linguistic interactions, as couplings of structured semiotic systems. The hidden mechanisms and regulatory interactions studied by Hofer can be treated as linguistic interactions between semiotic systems as well. This can explain how human relations have deep, complex, levels: symbolic, but also iconic and informational layers. These different layers are interacting and intermingled within each individual system (as in the classic body-mind framework) but also between individuals and groups. Overt verbal communication and its interpretations are only a partial area of the global semiotic life. This can be valid also in the psychotherapeutic setting, which can be regarded as a peculiar and partially controlled semiotic experience.

The amazing complexity of human semiotic life can be highlighted by some fragments of clinical sessions. For example, smell has an important and often hidden role. A patient, a young girl, used to express her emotions by bringing into my office bags full of vegetables or fruits with strong smells. When she was happy she brought oranges, when she was sad she brought onions. In this way she could occupy the available space in the room, also penetrating in my body by the air, and leave a marker of her passage in my office after the end of the session.

In another psychoanalytical treatment the sensory interaction was mediated by the sound of the voice and some qualities expressed in metaphors. The analysand was another young girl in severe depression. The psychoanalytical relationship seemed more based on sensory qualities (soft-hard; warm-cold etc.) than on explicit verbal communication. Several months after she had begun treatment she declared that often she listened only to the tone of the therapist's voice, and that it was the most important thing in analysis, to her. She said: "words, even my own, often seem made by metal or cardboard".

These clinical sketches suggest that there are interactions between the different layers of sensory and linguistic experience. The latter patient could reach a metaphorical expression of her sensations; while the former used fruits and vegetables both as a source of indexes and as symbols, following the culture bound meanings that we assign to them. The semantic and syntactic levels of our relationships are specific to being human, but the old Thomist statement that "the Word kills the Thing" seems a sin of intellectual arrogance: hidden dynamics of indexes and information are a basic part of our biopsychosocial interactions. As stated by Kelso "it seems justified to conclude that order parameters in biological systems are functionally specific, context sensitive informational variables; and that the coordination dynamics are more general than the structures that instantiate them"[89]. Self-organized coordination dynamics are, at their very roots, informational, and they are quite independent from the physical medium through which they are realized. Informational structures can make couplings between different physical media.

If speech and thought can be expressed, like biological and physical variables, in dynamic patterns of spatiotemporal activities, then there is no "mysterious leap between the mind and the body". Shared principles of self-organization provide the linkage across levels of neural and cognitive functions[90]. In this way we can come back to the original Es (Id)[91], whose name was derived by Groddeck from Nietzche as "...what there is of not-personal and, we can say, necessary by nature in our being".

If our examination of these levels of the psychobiological life is correct there are some clinical implications in the conduction of treatments. As Taylor[92] proposed in several contributions, the main aim of the psychosomatic approach should be to enhance the patient's self regulating capacities. Taylor suggests the traditional, "object relations", approach of "the resolution of interpersonal and intrapsychic conflicts, and the unification of split 'good' and 'bad' representations of self and object"[93]: " a lengthy process during which the therapist must function like a 'good enough mother', who contains and transforms her infant's primitive anxieties and facilitates the emergence of transitional activities and other self regulating capacities." The therapeutic task depends on the level, or the site, to which we assign the control of the self regulatory mechanisms. Even if we follow Hofer these mechanisms seem different from the traditional representation of *objects* in the psychoanalytic theory. The sensory objects defined by Hofer are not mental representations, but organized patterns of sensory dynamics.

The general metaphorical framework of the 'good enough mother' could be still valid, but our work could be aimed to change, by well-timed perturbations, the coupling between the active parts of the attractors and the patient's mind as a whole. We should make the patient's mind processes able to have larger dimensions, and being more flexible, facilitate continuous passages among indexes, icons and symbols, and their expression. This work can be partially under the conscious control of the therapist, but it is partially out of control, based on the random meeting of a patient's and an analyst's dimensions on the landscape of the treatment.

As we examined in the Behcet's case history, 'hidden factors' are an important part of determinant factors in the success or failure in the therapeutic process. They depend also of the (unconscious) random meeting of a particular therapist with a particular patient, and the singular hidden mechanisms of their interaction. Clearly, it is impossible to have a total control over the bulk of variables coupled in the interaction, but *Non Linear Dynamics* are giving us new ways to better understand the way we cure, and we may improve our methods.

8 Science and Metaphor in Mind Sciences

The morphological   approach could open new perspectives on the effective isomorphisms and couplings in different biological and cognitive domains of the human realm. It would not be as misleading as the reductionistic approach present, for instance, in some authors in the field of Neurosciences, where the discovery of microtubules and related quantistic effects in cellular protoplasm, leaded to the generalization of quantum dynamics in human thinking. This was both a form of reductionism, and the metaphorical use of a scientific discovery. A scientific find or a well known rule can be valid for one scale of events, but not for another. A generalization could be useful in a metaphorical usage, which should be overtly declared to avoid any equivocation.

Metaphor has been seen in the rethoric tradition as brief similarity (*similitudo brevior*). In metaphor the overlapping of a small part of the semantic fields of two different terms allow the use of one in the place of the other. The use of language is a basic part of our knowledge of the world and metaphor is a tool in the creative use of language. We may recall the well known example of the Eskimo names for 'snow' to demonstrate that words make us see things. Metaphor builds, through addictions and subtractions of semantic links, new areas of meaning and understanding of reality.

Sometimes, as in the so-called hermeneutics, this power of the metaphorical process is used for endless fugues in the linguistic universe. This use of metaphor is closer to the arts, for instance poetry where reaches aesthetic and creative results, than to science. It is a logical consequence of the geometries of language as highlighted by Peirce's definition of meaning as 'interpretant'.

Max Black[94] asserted that metaphors are to literature what models are to science. Metaphors are a creative second thought attempting to "cut the world" in a new way[95,96]. The passage from mythological to scientific thinking is related to the stabilization of a metaphor in the form of a model or a rule, maintaining robust correlations with experience, falsificability, inner coherence and verifiability.

"Metaphoric uses sometimes can verbalize subtleties that mathematical modelling might overlook. Metaphoric uses often imply that while we may not have formalized a math model, the parsimony and power of dynamical processes obeying the usual mathematical properties but yet unexpressed may be lurking as driving the processes we describe metaphorically. That is the insights of the math approaches empower the metaphoric. Both of these ideas suggest a powerful synergy"[97].

Unfortunately, sometimes, metaphors are crystallized in metaphysics. Psychoanalytic practice has not degenerated in a sort of sophistry, an art of persuasion, because it remains strictly linked to some basic cores. In another context I defined them as *"pre-existing metapsycological cores"*[98], which denounce an evident metaphysical nature. They are part of the area of values that include ethics, religion, and sometimes magic[99,100,101]. It is not a casual fact that Freud himself declared to be spellbound by "metapsychology".

"We can only say: 'So muss denn doch die Hexe dran ! ('We must call the Witch to our help after all !' - Goethe, Faust, part 1, scene 6) - the Witch Metapsychology. Without metapsychological speculation and theorizing - I had almost said 'phantasying' - we shall not get another step forward. Unfortunately, here as elsewhere, what our Witch reveals is neither very clear nor very detailed"[102].

We might consider the psychoanalytical definition of what could be seen as a primary state of "non-unity" of the mind. Freud calls it "auto-erotism"; Glover writes about "Ego nuclei"; Winnicott defines it as a "primary state of non-integration"; M. Klein suggests an Ego disintegration under the inner impact of death instincts; Fairbairn asserts that the 'originary unity' is followed by the disintegration caused by "internalized external persecutory objects"; Lacan points out that the original subject has been "barred" and excluded from full enjoyment; in Bion's writings it is the "absent object" as a "non-object" to establish the foundations of thinking. And the list could go on.

These are genuine myths of the Subject's origins, a sort of *"Subject's cosmogonies"*. They warp the virtual semiotic landscapes of each treatment; they orient the paths of each psychoanalysis; they assign a special value to every event during the course of the cure. This *"metaphysical power"* of psychoanalytical models has been a barrier against sophist drifts. But it has also been a strong source of stagnation in research.

We must specify that the use of the term *'model'* in the psychoanalytical field is particular, and different from most fields of science. Usually, a model is "a schematic presentation of reality connecting the main quantities by laws which take the form of mathematical equations"[103]. It is obvious that what have been called 'models' in psychoanalysis are for the most part metaphors and rarely topological attempts. Mathematics and quantities are not implemented in psychoanalytical studies: there are only a few peculiar exceptions.

NonLinear Dynamics provide us with a number of new metaphors such as self-organization, strange attractors, degrees of freedom etc. which may play a part in all kinds of disciplines. These metaphors are closer to the features of the fields of *Mind Sciences* and could be bridges to allow a transition to a more appropriate and non-reductionistic mathematical and quantitative approach: towards science, leaving metaphysics to the areas of ethics and religion.

References

1. S. Freud, *Abhandlungen an Wilhelm Fliess/ Letters to W.F.*, (London UK,Imago,Standard Ed. vol. 1,1895).
2. C.S. Peirce, *Letters to Lady Welby*, (New York USA,Whitlock's, 1908 - Lieb IC ed.,1953).
3. F. Alexander, *Psychosomatic Medicine*, (New York, Norton, 1950).
4. F. Dunbar, *Mind and Body: Psychosomatic Medicine*, (New York,Random House, 1947).
5. R.K. Eissler, The effect of the structure of the ego on psychoanalytic technique, J.Am.Psychoan.Assn. **1** (1953).
6. S. Freud, *Das Ich und das Es/The Ego and the Id.* (London,Imago,Standard Edition,vol.**19**:3, 1922-1946).
7. G.L. Engel, The clinical aplication of the biopsychosocial model. Am.J.Psychiat. **137**:535-544 (1980).
8. C.P. Kimball, *The Biopsychosocial Approach to the Patient.* (Baltimore,Williams & Wilkins,1981).
9. Z.J. Lipowski, What does the word "psychosomatic" really mean ? Psychosom.Med. 46:153-171,1984.
10. G. Groddeck, *Das Buch vom Es.* (Wien,Deuticke,1923, en.ed.New York,Vintage, 1949).
11. F. Orsucci, Behcet's Disease and psychosomatic patterns of thinking. Psychother.Psychosom. **65**:112-114,1996.
12. G.R. Plotkin et al.(eds), *Behcet's disease: a contemporary synopsis.* (Mount Kisco, N.Y., Futura Pub. Co,1988).
13. G . Inaba(ed), *Behcet's disease: pathogenetic mechanism and clinical future.*, (Tokyo ,University of Tokyo Press,1982).
14. J. Baticle, *Goya d'Or et de Sang.*, (Paris,Gallimard,1986).
15. S. Marvin, Stress, depression and the immune system. J Clin Psychiatry **50**; 35-40,1989.
16. M.R. Irwin & H. Strausbaugh,Stress and immunechanges in human: a biopsycosocial model, Prog Psychiatry **35**; 56-80,1992.
17. N.P. Plotnikoff et al., *Stress and Immunity.* (Ann Arbor-London, CRC Press,1991).
18. G.J. Taylor, Alexithymia: Concept, measurement and implication for treatment.

Am J Psychiatry **141**; 725-752,1984.

19. P.W. Linville, Self-complexity: a cognitive buffer against stress-related illness. J Person Soc Psychol **52** (4):663-680,1987.

20. M.A. Hofer, Relationships as regulators. Psychosom Med **46**:183-197,1984.

21. S. Freud,*Entwurf eine Psychologie./ Project for a Scientific Psychology.*, (London UK,Imago,Standard Ed. vol. **1**, 1895-1946).

22. J. Lacan, *Ecrits.* (Paris F,Editions du Seuil,1966).

23. S. Freud, *Hemmung, Symptom und Angst / Inhibition,Symptom and Anxiety.* (London UK,Imago,Standard Edition,vol.**20**:77,1925-1946).

24. W.R. Bion, *Second Thoughts. (Selected Papers of Psychoanalysis).* (London UK,Heinemann,1967).

25. W.R. Bion, *Cogitations.* (London,Karnak Books,1992).

26. U. Eco,*A Theory of Semiotics.*, (Bloomington,Indiana Univ.Press,1975).

27. W.J. Freeman,*Societies of Brains.*, Hillsdale NJ,Lawrence Erlbaum,1995.

28. W.J., Freeman *Objective self, subjective self, and the illusion of control.*, (unpublished manuscript, 1996).

29. W.R. Fairbairn, *Psychoanalytic Studies of the Personality.*, (London,Routledge,1952).

30. F. Orsucci et al, Alexitima: espressione dele emozioni e funzioni simboliche. Med.Psicosom. **38**:225-232,1993.

31. G.J. Taylor et al., Toronto Alexithymia Scale, Psychother.Psychosom. **57**:34-41, 1986.

32. H. Krystal et al, Assessment of alexithymia in posttraumatic stress disoder, Psychosom.Med. **48**:89-94,1986.

33. G. Mendelson, Alexithymia and chronic pain. Psychother.Psychosom. **18**:363-373,1982.

34. P.E. Gahrnkel,D.M. Gardner,*Anorexia Nervosa: A Multidimensional Perspective.* New York,Brunner-Mazel,1982.

35. J.J. Lopez Ibor, Masked Depressions. Brit.J.Psychiat. **120**:245-258,1972.

36. J. McDougall, Alexithymia: a psychoanalytic viewpoint
Psychother.Psychosom. **38**:81-90,1982.

37. T. Wise et al, Secondary alexithymia: an empyrical validation. Comp.Psychiat. **31**:284-288,1990.

38. J. TenHouten et al, Alexithymia and the split brain.
in K.D. Hoppe(ed) *The Psychiatric Clinics of North America.*, Philadelphia,Saunders,1988.

39. J. Zeitlin et al, Interemispheric transfer deficit and alexithymia.

Am.J.Psychiat. **146**:1434-39,1989.

40. K.K.S.Voeller, Right-emisphere deficit syndrome in children.
Am.J.Psychiat. **143**:1004-09, 1986.

41. J. Parker et al, Relationship between lateral conjugate eye movements and alexithymia, Psychother.Psychosom. **57**:94-101,1992.

42. P. Marty,M. De M'Uzan, La "pensée operatoire".
Rev.Franç.Psychanal.**27**:1345-56, 1963.

43. J. Piaget, *Le Langage et la Pensée chez l'Enfant.*
Paris,Delachaux et Niestlé, 1923.

44. J. Piaget, *Six études de Psychologie.*
Paris,Gonthier ,1964.

45. R.D. Lane,G.E. Schwartz, Levels of emotional awareness.
Am.J.Psychiat. **144**:133-143,1987.

46. Freud S, *Entwurf eine Psychologie./ Project for a Scientific Psychology.*, (London UK,Imago,Standard Ed. vol. **1**, 1895-1946).

47. G.T. Fechner, *Einige Ideen zur Schopfungs- und Entwicklungs-geschichte.*
Leipzig,Breitkopf und Härtel, 1873.

48. S. Freud, *Jenseits der Lustprinzip/Beyond the Pleasure Principle.*
London UK,Imago,Standard Edition, vol.18:7, 1920.

49. W.B. Cannon, *The Wisdom of the Body.*
New York,Norton, 1939.

50. G.L. Engel,A.H. Schmale, Psychoanalytic theory of somatic disorder.
J.Am.Psychoan.Assn. **15**:344-365, 1967.

51. H. Weiner, The prospects for psychosomatic medicine.
Psychosom.Med. **44**:491-517, 1982.

52. M.A. Hofer, Hidden regulatory processes in early social relationships.
In:*Perspectives in Ethology*,Bateson PG & Klopfer PH(eds.),New York,Plenum, 1978.

53. G.J. Taylor, *Psychosomatic Medicine and Contemporary Psychoanalysis.*
NewYork,International Universities Press, 1987.

54. A.L. Goldberger et al, On a mechanism of cardiac electrical stability.
Biophys.J. **48**:525, 1985.

55. A.L. Goldberg,B.J. West, Chaos and order in the human body.
Science, **9/1**:25-34 , 1992.

56. A.L. Goldberger,B.J. West, Applications of nonlinear dynamics to clinical cardiology.
Ann.N.Y.Acad.Sci.,**504**:195, 1987.

84

57. R. Sinha et al, Cardiovascular differentiation of emotions.
Psychosom.Med. **54**:422-435, 1992.

58. C.A. Skarda,W.J. Freeman, How brains make chaos in order to make sense of the world. Brain & Behavioral Science **10**:161-195, 1987.

59. J.H. Are,R.H. Simon, An application of the fractal dimension.
Electronceph.Clin.Neurophysiol. **75**:296-305, 1990.

60. L.D. Iasemedis et al, Phase space tomography and the Lyapunov exponent.
Brain Topog. **2**:187-201. 1990.

61. A. Babloyantz,A. Destexhe, Low dimensional chaos in an instance of epilepsy.
Proc.Natl.Acad.Sci.USA **83**:3513, 1986.

62. S.J. Schiff et al, Controlling chaos in the brain.
Nature **370**:615-620, 1994.

63. J. Glanz, Do chaos control techniques offer hope for epilepsy ?
Science **265**:1174, 1994.

64. R.A. Spitz, Hospitalism. Psychoan.Study.Child **1**:53-74, 1945.

65. G.L. Engel,A.H. S chmale, *Conservation withdrawal:A primary regulatory process*. Amsterdam,Ciba Foundation Symposium 8,Elsevier, 1972.

66. M. Maes et al., Autoimmunity and depression.
Acta Psychiat.Scand. **87**:160-166, 1993.

67. J. Bowlby, *Attachment and Loss, Vol.1 Attachment*.
New York,Basic Books, 1969.

68. J. Bowlby, *Attachment and Loss, Vol.2 Separation*.
New York,Basic Books, 1973.

69. M.A. Hofer, *The Roots of Human Behavior*.
San Francisco,Freeman, 1981.

70. G.E. Evoniuk et al, The effect of tactile stimulation on serum growth hormone.
Communicat.Psychopharmacol. **3**:363-370, 1979.

71. E.A. Stone et al, Survival and development of maternally deprived rats.
Psychosom.Med. **38**:242-249, 1976.

72. M. Mahler, *On Human Symbiosys and the Vicissitudes of the Individuation.*,
New York, International Univ. Press 1968.

73. J. Bowlby, *Attachment and Loss, Vol.1 Attachment*.
New York,Basic Books 1969.

74. J. Bowlby, *Attachment and Loss, Vol.2 Separation*.
New York,Basic Books, 1973.

75. D.W.Winnicott, *The Maturational Processes and the Facilitating Environment.*,
London,Hogarth Press, 1965.

76. H. Kohut, *How Does Analysis Cure ?*
    Chicago,University of Chicago Press, 1984.
77. C. Ehlers, Chaos and complexity. Can it help us to understand mood and
    behavior ? Arch Gen Psychiat **52**:960-964, 1995.
78. C. Ehlers et al, Social zeitgebers and biological rythms.
    Arch Gen Psychiat **45**:948-952, 1988.
79. C.E. Shannon, Prediction and entropy of printed English.
    Bell Syst Tech J. **30**:50, 1951.
80. P. Grassberger, Estimating the information content of symbol sequences and
    efficient codes. IEEE Trans Inf Theory,**35**:669, 1989.
81. Y.C. Zhang, Complexity and 1/f noise: a phase space approach.
    J Phys (France), **11**:971, 1991.
82. A. Schenkel et al, Long range correlation in human writings.
    Fractals,**1-1**:47-57, 1993.
83. M. Amit et al, Language and codification dependence of long range correlations
    in texts. Fractals,**2-1**:7-13, 1993.
84. F. Orsucci et al, Caos e morfogenesi nelle dinamiche nonlineari della mente.
    In:Liotta E,Orsucci F,Turno M(eds.), *Between Order and Chaos*,Rome,Teda,
    1996.
85. F. Orsucci & A. Giuliani, The orthographic-stylistic structuring of human speeches.
    Proceedings Workshop *"Complexity in the Living"*,Rome, in press.
86. J.P. Eckman,S.O. Kamphorst & D. Ruelle, Recurrence plots of dynamical systems.
    Europhys Lett,**4**:973-977, 1987.
87. C.L. Webber and J.P. Zbilut, Dynamical assessment in physiological systems
    and states. J.Appl.Physiol. **76**:965-973, 1994
88. J.S. Nicolis & A.A. Katsikas, Chaotic dynamics of linguistic-like processes.
    In,West BJ (ed),*Patterns Information and Chaos in Neuronal
    Systems*,Singapore,World Scientific Publishing, 1993.
89. J.A.S. Kelso, *Dynamic Patterns, The Self-Organization of Brain and Behavior.*
    Cambridge Mass.,MIT Press, 1995.
90. H. Atlan, Self-creation of meaning. Physica Scripta **36/2**:563-576, 1987.
91. Freud S, *Das Ich und das Es/The Ego and the Id.*
    (London,Imago,Standard Edition,vol.**19**:3, 1922-1946).
92. G.J. Taylor, Psychoanalysis and empyrical research.
    J.Am.Acad.Psychoanal. 23:263-281, 1995.
93. G.J. Taylor, Psychoanalysis and psychosomatics: a new synthesis.
    J.Am.Acad.Psychoanal. 20:251-275, 1992.

94. M. Black, *Models and Metaphors.*
Ithaca,Cornell Univ.Press, 1962.

95. R. Boyd, Metaphor and theory change: what is 'Metaphor' a metaphor for ?
In:*Metaphor and Thought* (Orthony A.ed),Cambridge UK,Cambridge Univ.
Press, 1979.

96. T.S. Kuhn, Metaphor in science.
in *Metaphor and Thought* (Orthony A.ed),Cambridge UK,Cambridge Univ.
Press, 1979.

97. F. Abraham, Contribution of February 15.
The Dialogue Discussion Group,Internet, 1997.

98. F. Orsucci, Introduzione. In:Fairbairn WR,*Scritti 1952-1963*,Roma,Astrolabio-
Ubaldini, 1993.

99. L. Wittgenstein,*Philosophische Untersuchungen / Phylosophical Investigations..*
Oxford UK, Basil Blackwell, 1953.

100. L. Wittgenstein, *Bemerkungen über Frazer's 'The Golden Bough".*
Dordrecht,Synthese, 1967.

101. L. Wittgenstein, *Remarks on the Philosophy of Psychology.*
Oxford,Basil Blackwell, 1980.

102. Freud S *Die endliche und die unendliche Analyse / Analysis Terminable and.
Unterminable.* London UK,Imago,Standard Ed. vol.23, 1937.

103. F. Verhulst, Metaphors for psychoanalysis.
Nonlinear Science Today **4/1**:1-6, 1994.

MINIMAL MODELS FOR DYADIC PROCESSES: A REVIEW

SERGIO RINALDI

Sergio Rinaldi, Dipartimento di Elettronica e Informazione, Politecnico di Milano, Via Ponzio 34/5, 20133 Milano, Italy. E-mail: rinaldi@elet.polimi.it

ALESSANDRA GRAGNANI

Alessandra Gragnani, Dipartimento di Elettronica e Informazione, Politecnico di Milano, Via Ponzio 34/5, 20133 Milano, Italy. E-mail: gragnani@elet.polimi.it

This paper is a survey of a few recent contributions in which dyadic processes are studied as formal dynamical systems. For this, a general minimal model composed of two ordinary differential equations is first considered as a possible formal tool to mimic the dynamics of the feelings between two persons. The equations take into account three mechanisms of love growth and decay: the pleasure of being loved (return), the reaction to partner's appeal (instinct), and the forgetting process (oblivion). Under extremely simple assumptions on the behavior of the individuals, the minimal model turns out to be a positive linear system enjoying, as such, a number of remarkable properties, which are in agreement with common wisdom on the argument. These properties are used to explore the consequences that individual behavior can have on community structure. The main result along this line is that individual appeal is the driving force that creates order in the community. Then, in order to make the assumptions more realistic, in accordance with attachment theory, individuals are divided into secure and non secure individuals, and into synergic and non synergic individuals, for a total of four different classes. Using always the same minimal model, it is shown that couples composed of secure individuals, as well as couples composed of non synergic individuals can only have stationary modes of behavior. By contrast, couples composed of a secure and synergic individual and a non secure and non synergic individual can experience cyclic dynamics. In other words, the coexistence of insecurity and synergism in the couple is the minimum ingredient for cyclic love dynamics. Finally, a slightly more complex model, composed of three ordinary differential equations, proposed to study the dynamics of love between Petrarch, a celebrated Italian poet of the 14-*th* century, and Laura, a beautiful but married lady, is also reviewed. Possible extensions are mentioned at the end of the paper.

1 Introduction

This paper deals with *love dynamics*, a subject which falls in the field of social psychology, where interpersonal relationships are the topic of major concern. Romantic relationships are somehow the most simple case since they involve only two individuals.[1]

Love-stories are dynamic processes that start from zero (two persons are completely indifferent one to each other when they first meet), develop (more or less quickly) and end up into some sort of regime. Real-life observations tell us that most of the times transients develop quite regularly and asymptotic regimes are stationary and associated to positive romantic relationships. But there are also love-stories initially characterized by stormy patterns of the feelings as well as by cyclic regimes, like that identified by Jones[2] in Petrarch's *Canzoniere*, the most celebrated book of love poems of the Western world.

What has been said about steady and cyclic romantic relationships reminds very much the behavior of dynamical systems tending toward equilibria or limit cycles. But observations also point out the existence of multiple attractors. For example, it is known that steady and high quality romantic relationships can turn into a state of permanent antagonism after a disturbance, for example after a temporary infatuation of one of the two partners for another person. The above remarks spontaneously suggest the use of differential equations for modelling the dynamics of the feelings between two individuals.

Although dynamic phenomena in physics, chemistry, economics and all other sciences have been extensively studied by means of differential equations, surprisingly such an approach has never been followed in social psychology. The first contribution along this line are two unpublished notes by Etienne Guyon and Steven Strogatz, written in 1975 and 1978, respectively, in which a very simple linear model for a generic couple is analyzed. The first published contribution is a one page paper in which Strogatz[3] describes how the classical topic of harmonic oscillations can be taught to capture the attention of students. He suggests to make reference to "a topic that is already on the minds of many college students: the time evolution of a love affair between two people". The model proposed by Strogatz (discussed also in Radzicki[4] and Strogatz[5]) is definitely unrealistic because it does not take into account the appeal of the two individuals. Thus, Strogatz's model does not explain, for example, why two persons who are initially completely indifferent one to each other can develop a love affair.

More realistic models have been proposed and studied by the authors in the last years. Three aspects of love dynamics are taken into account in these models: the forgetting process, the pleasure of being loved, and the reaction to the appeal of the partner. In a first model, proposed by Rinaldi,[6] these three factors are modelled by linear functions. The resulting model is a linear dynamical system, which turns out to be positive if the appeals of the two individuals are positive. The theory of positive linear systems[7,11] can therefore be applied to this model and gives quite interesting results. Some of them describe the dynamic process of falling in love, *i.e.* the transformation of the feelings, starting from complete indifference (when two persons first meet) and tending toward a plateau. Other results are concerned with the influence that appeal and individual behavior have on the quality of romantic relationships. Some of these properties are used to identify the consequences that individual appeal and behavior can have on partner choice and on community structure. Although the results are extreme, they explain to some extent facts observed in real life, such as the rarity of couples composed of individuals with very uneven appeal.

In a second model, proposed by Rinaldi and Gragnani,[12] the pleasure of being loved and the reaction to the appeal are modelled by nonlinear functions taking into account the traits of so-called secure individuals.[13,14] This model retains many of the properties of the simpler linear model and allows one to derive the same conclusions at individual and community level.

In a third model, proposed by Gragnani et al.,[15] synergism, i.e. the fact that an individual may react more strongly when he is in love, has also been modelled. The result is that the coexistence of unsecurity and synergism in a couple is the reason for cyclic romantic relationships.

There is, finally, a very recent study[16] dealing with a well documented and particular case: the dynamics of love between Petrarch, an Italian poet of the 14-th century, and Laura, a beautiful and married lady. Their love story developed over 21 years and has been described in the Canzoniere, a collection of 366 poems addressed by the poet to his platonic mistress. In such a study, the main traits of Petrarch's and Laura's personalities are identified by analyzing the Canzoniere. These traits are encapsulated in a model composed by differential equations where the variables are the emotions of the two individuals. The model is analyzed and the result is that its solution tends toward a cyclic behavior. In other words, the peculiarities of the personalities of the two lovers inevitably generate a never ending story of recurrent periods of ecstasy and despair. This is indeed what happened, as empirically ascertained by Jones[2] through a detailed stylistic and linguistic analysis of the dated poems. It is interesting to note, however, that with the modelling approach the existence of the emotional cycle is fully understood and proved to be inevitable, while with the empirical approach it is only discovered.

Aim of this paper is to review the main results and properties of the above models, each one considered as a particular case within a general class of minimal models. The paper is organized as follows. In the next section we describe the structure of a minimal model. Then we identify the models described by Rinaldi,[6] Rinaldi and Gragnani[12] and Gragnani et al.[15] as particular cases within the class of minimal models and study their properties. Finally we present the study on Laura and Petrarch. Possible extensions are briefly discussed at the end of the paper.

2. Structure of the minimal model

The model discussed in this section is a *minimal* model, in the sense that it has the lowest possible number of state variables, namely one for each member of the couple. Such variables, indicated by x_1 and x_2, are a measure of the love of individual 1 and 2 for the partner. Positive values of x represent positive feelings, ranging from friendship to passion, while negative values are associated with antagonism and disdain. Complete indifference is identified by $x = 0$.

The model is a crude simplification of reality. Firstly, because love is a complex mixture of different feelings (esteem, friendship, sexual satisfaction, ...) and can be hardly captured by a single variable. Secondly, because the tensions and emotions involved in the social life of a person cannot be included in a minimal model. In other words, only the interactions between the two individuals are taken into account, while the rest of the world is kept frozen and does not participate explicitly in the formation of love dynamics.

Three basic processes, namely, *oblivion*, *return* and *instinct*, are assumed to be responsible of love dynamics. More precisely, the instantaneous rate of change $dx_i(t)/dt$ of individual's i love is assumed to be composed of three terms, *i.e.*,

$$\frac{dx_i}{dt} = O_i + R_i + I_i \tag{1}$$

where O_i, R_i and I_i must be further specified.

The oblivion process can be easily studied by looking at the extreme case of an individual who has lost the partner (which implies $R_i = I_i = 0$). If we assume that in such conditions $x_i(t)$ vanishes exponentially at a rate α_i, we must write

$$\frac{dx_i(t)}{dt} = -\alpha_i x_i(t)$$

so that we can derive

$$O_i = -\alpha_i x_i(t)$$

The return R_i describes the reaction of individual i to the partner's love and can therefore be assumed to depend upon x_j, $j \neq i$. In order to be more specific on the return function R_i we subdivide individuals into two classes: secure and non secure individuals. Secure individuals have positive mental models of themselves and of the others and their romantic relationships are characterized by intimacy, closeness, mutual respect and involvement. They react positively to partner's love and are not afraid about someone becoming emotionally close to them. Therefore, we could assume that in the extreme case the return function R_i is linearly dependent upon x_j. But more realistically, we must assume that R_i is positive [negative], increasing, concave [convex] and bounded for positive [negative] values of x_j. The boundedness of the return function is a property that holds also for non secure individuals: it interprets the psycho-physical mechanisms that prevent people from reaching dangerously high stresses. By contrast, since non secure individuals react negatively to too high pressures and involvement,[14] their return function R_i is decreasing for high values of x_j. Figure 1a shows the graphs of typical return functions R_i for secure and non secure individuals.

The third term I_i describes the reaction of individual i to the partner's appeal A_j. Of course, it must be understood that appeal is not mere physical attractiveness, but, more properly and in accordance with evolutionary theory, a suitable combination of different attributes among which age, education, earning potential and social position. Moreover, there might be gender differences in the relative weights of the combination.[17,18] For reasons similar to those mentioned above, we assume that instinct functions $I_i(A_j)$ enjoy the same properties that hold for return functions for secure individuals. Nevertheless, for simplicity, we will restrict our attention to the case of individuals with positive appeals. Figure 1b shows the graphs of two typical instinct functions I_i.

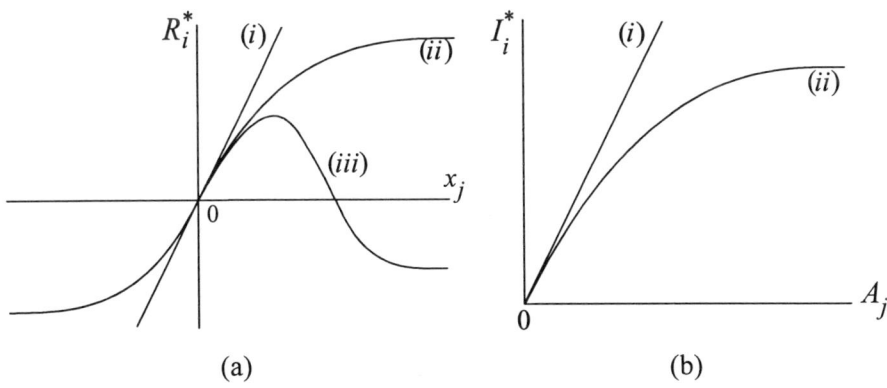

Figure 1 : (a) return functions $R_i^*(x_j)$ of secure individuals (linear (*i*) and bounded (*ii*)) and non secure individuals (*iii*); (b) instinct functions $I_i^*(A_j)$ (linear (*i*) and bounded (*ii*)).

Up to now, each individual i is identified in the model by two parameters (the appeal A_i and the forgetting coefficient α_i) and two functions (the return function R_i and the instinct function I_i). Such parameters and functions are assumed to be constant in time: this rules out aging, learning and adaptation processes which are often important over a long range of time[19,20] and sometimes even over relatively short periods of time.[21]

Moreover, it is known that individual reactions can be enhanced by love. For example, mothers have often a biased view of the beauty of their children. This kind of phenomenon, here called *synergism*, has been empirically observed in a study on perception of physical attractiveness[22] by comparing individuals involved in dating relationships with individuals not involved in them. Although we are not aware of any study pointing out the existence of synergism in the reaction to partner's love, we can reasonably assume that also return functions can be enhanced by love. Thus, we can consider reaction and instinct functions of the form

$$R_i = (1 + S_i^R(x_i))R_i^*(x_j)$$
$$I_i = (1 + S_i^I(x_i))I_i^*(A_j)$$

where the functions R_i and I_i (shown in Fig. 1) are, by definition, the reactions of a completely indifferent individual (*i.e.* the reaction of individual i with $x_i = 0$) and the functions $S_i^R(x_i)$ and $S_i^I(x_i)$ are zero for $x_i \leq 0$ and increasing, convex/concave and bounded for $x_i > 0$, as shown in Fig. 2. The upper bounds of the functions S_i^R and S_i^I are called *synergism coefficients*.

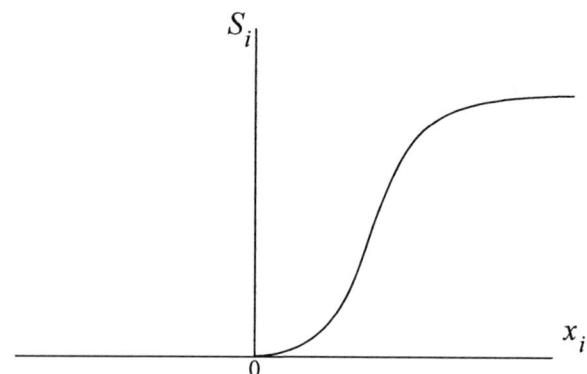

Figure 2 : The graph of a typical synergism function $S_i(x_i)$.

Individuals are therefore divided not only into secure and non secure individuals but also in synergic and non synergic individuals for a total of four classes. Notice that we have proposed a minimal model in which each individual is characterized by only one state variable. Nevertheless, we will see in the following that individuals with complex personalities may require more than one state variable (see Rinaldi[16]).

3. A simple linear model

In the model analyzed in this section, synergism is absent and all other processes are assumed to be linear so that

$$O_i = -\alpha_i x_i \qquad\qquad R_i = \beta_i x_j \qquad\qquad I_i = \gamma_i A_j$$

where α_i, β_i and γ_i, as well as the appeals A_i $(i = 1, 2)$, are constant and positive parameters.

The result is the following model

$$\dot{x}_1(t) = -\alpha_1 x_1(t) + \beta_1 x_2(t) + \gamma_1 A_2$$
$$\dot{x}_2(t) = -\alpha_2 x_2(t) + \beta_2 x_1(t) + \gamma_2 A_1 \tag{2}$$

In the first model discussed and published in the literature[3] (see also Radzicki[4] and Strogatz[5]) oblivion and instinct are neglected ($\alpha_i = \gamma_i = 0$), while the two coefficients (β_i) characterizing the return have opposite sign, i.e., one of the two lovers is a bit masochist and hates to be loved and loves to be hated.

Model '(2) is a linear system which can be written in the standard form $\dot{x} = Ax + bu$ with $u = 1$, and

$$A = \begin{vmatrix} -\alpha_1 & \beta_1 \\ \beta_2 & -\alpha_2 \end{vmatrix} \qquad\qquad b = \begin{vmatrix} \gamma_1 A_2 \\ \gamma_2 A_1 \end{vmatrix}$$

Such a system is positive because the matrix A is a Metzler matrix (non negative off-diagonal elements) and the vector b has positive components.[8] Thus, $x(0) \geq 0$ implies $x(t) \geq 0 \; \forall \; t$. This means that the assumptions imply that two persons of this kind will never become antagonist, because they are completely indifferent one to each other when they meet for the first time (i.e. $x(0) = 0$).

Positive linear systems enjoy remarkable properties, if they are asymptotically stable. In the present case, the necessary and sufficient condition for asymptotic stability is $\beta_1\beta_2 < \alpha_1\alpha_2$ *i.e.* the system is asymptotically stable if and only if the (geometric) mean reactiveness to love ($\sqrt{\beta_1\beta_2}$) is smaller than the (geometric) mean forgetting coefficient ($\sqrt{\alpha_1\alpha_2}$). In the following, the above condition is assumed to hold. Thus, system (2) is asymptotically stable and the love of each individual is bounded and tends toward an equilibrium value $\bar{x}_i$, which must be non-negative because the system is positive. We now state five simple but interesting properties of model (2) (for the proofs of these properties, the interested reader can refer to Rinaldi[6]).

Property 1
The equilibrium $(\bar{x}_1,\bar{x}_2)$ of system (2) is strictly positive.

Property 2
The function $x_i(t)$, corresponding to the initial condition $x(0) = 0$, is strictly increasing.

Property 3
An increase of the reactiveness to love [appeal] β_i [γ_i] of individual i gives rise to an increase of the love of both individuals at equilibrium. Moreover, the relative increase is higher for individual i.

Property 4
An increase of the appeal A_i of individual i gives rise to an increase of the love of both individuals at equilibrium. Moreover, the relative increase is higher for the partner of individual i.

Property 5
An increase of the reactiveness to love gives rise to an increase of the dominant time constant of the system.

The above five remarks can be easily interpreted. The first states that individuals with positive appeal are capable of establishing a steady romantic relationship. The emotional pattern of two persons falling in love is very regular - beginning with complete indifference, then growing continuously until a plateau is reached (Property 2). The level of passion characterizing this plateau is higher in couples with higher reactiveness and appeal (Properties 3, 4). Moreover, an increase in the reactiveness of one of the two individuals is more rewarding for the same individual, while an increase of the appeal is more rewarding for the partner. Thus, there is a touch of altruism in a woman [man] who tries to improve her [his] appeal.

Finally, couples with very high reactiveness respond promptly during the first phase of their romantic relationship, but are very slow in reaching their plateau (Property 5). Together with Property 3, this means that there is a positive correlation between the time needed to reach the equilibrium and the final quality ($\bar{x}_1$ and $\bar{x}_2$) of the relationship. Thus, passions that develop too quickly should be expected to be associated with poor romantic relationships.

Some of these properties, which are at "individual" level, can be used to infer properties at "community" level. To show this, let us consider a community composed of N linear couples and let us say that this community is *unstable* if a woman and a man of two distinct couples believe they could be personally advantaged by forming a new couple together. In the opposite case the community is *stable*. Obviously, this definition must be further specified. The most natural way is to assume that individual i would have a real advantage in changing the partner, if this change is accompanied by an increase of $\bar{x}_i$. However, in order to forecast the value $\bar{x}_1$ [$\bar{x}_2$] that a woman [man] will reach by forming a couple with a new partner, she [he] should know everything about him [her]. Generally, this is not the case and the forecast is performed with limited information. Here we assume that the only available information is the appeal of the potential future partner and that the forecast is performed by imagining that the behavioral parameters of the future partner are the same as those of the present partner. This choice obviously emphasizes the role of appeal, quite reasonably, however, because appeal is the only easily identifiable parameter in real life. Within this framework it can be proved (see Rinaldi[6]) that stable communities are characterized by the following very simple but peculiar property involving only appeal.

Property 6
A community is stable if and only if the partner of the *n-th* most attractive woman ($n = 1, 2, ..., N$) is the *n-th* most attractive man.

This means that individual appeal is the driving force that creates order in our societies. This result, derived from purely theoretical arguments, is certainly in agreement with empirical evidence. Indeed, partners with very uneven appeals are rarely observed in relatively stable communities. Of course, in making these observations one must keep in mind that appeal is an aggregated measure of many different factors (physical, financial, intellectual, ...). Thus, for example, the existence of couples composed of a beautiful lady and an unpleasant but rich man does not contradict the theory, but, instead, confirms a classical stereotype.

4. More realistic nonlinear models

In the previous section, it has been shown that the simple linear model is asymptotically stable when the two individuals are not too reactive to partner's love. Under this assumption, the couple can reach a steady and positive romantic relationship. But, when this is not the case, the instability of the model gives rise to unbounded feelings, a feature which is obviously unrealistic. In such a case one must model the couple more carefully by assuming, for example, that the reaction function is increasing but bounded with respect to partner's love (see curve (ii) in Fig. 1a). This and other extensions will be considered in this section.

4.1 Secure - non synergic couples

In this subsection we refer again to couples composed of secure and non synergic individuals[12] ($i.e.$ return R_i and instinct I_i are still assumed to depend only upon x_j and A_j, respectively), but the dependence is nonlinear. The model is therefore

$$\dot{x}_1 = -\alpha_1 x_1 + R_1^*(x_2) + I_1^*(A_2)$$

$$\dot{x}_2 = -\alpha_2 x_2 + R_1^*(x_1) + I_2^*(A_1)$$

(3)

where the graphs of the functions R^* and I^* are like in Figs. 1a (graph (ii)) and 1b (graph (ii)).

Limit cycles can not exist in model (3). In fact, the divergence of the system (equal to $-(\alpha_1 + \alpha_2)$) does not change sign and Bendixon's criterion implies the non-existence of limit cycles. Moreover, the model is positive, since if the feelings are non-negative at a given time, then they are positive at any future time. This property implies that an undisturbed love story between secure individuals can never enter a phase of antagonism, a fact that is obviously against observations. Indeed, in real life the feelings between two persons are also influenced by facts that are not taken into account in the model. Of course, heavy disturbances can imply negative values of the feelings. It is therefore of interest to know what happens in such cases after the disturbance has ceased. The answer is given by the following property (proved in Rinaldi and Gragnani[12]).

Property 1'
Couples composed of secure and non synergic individuals can be partitioned into

robust and fragile couples. As time goes on, the feelings $x_1(t)$ and $x_2(t)$ of the individuals forming a robust couple tend toward two constant positive values no matter what the initial conditions are. By contrast, in fragile couples, the feelings evolve toward two positive values only if the initial conditions are not too negative and toward two negative values, otherwise.

Figure 3 illustrates Property 1' by showing in state space the trajectories of system (3) starting from various initial conditions. Note that no trajectory leaves the first quadrant. In Fig. 3a, corresponding to robust couples, all trajectories tend toward point $E^+ = (x_1^+, x_2^+)$, which is therefore a globally attracting equilibrium point. By contrast, fragile couples (Fig. 3b) have two alternative attractors (points E^+ and E^-) with basins of attraction delimited by the dotted line.

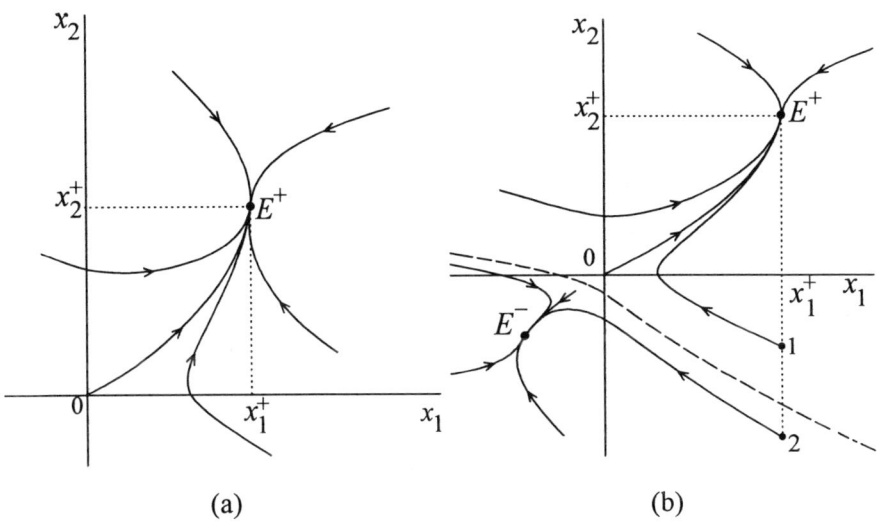

(a)   (b)

Figure 3 : Evolution of the feelings in a robust couple (a) and in a fragile couple (b) for different initial conditions. In (b) the boundary of the two basins of attraction (dashed line) is the stable manifold of a saddle.

Figure 3 also shows that the trajectory starting from the origin tends in any case toward the strictly positive equilibrium point E^+ without spiraling around it. This is equivalent to Properties 1 and 2 of the simple linear model. The above Properties 3 and 4, as well as Property 6 dealing with stable communities, are also retained by this model (see Rinaldi and Gragnani[12]).

Figure 3b holding for fragile couples points out an interesting fact. Suppose that the couple is at the positive equilibrium E^+, and that for some reason individual 2 has a drop in interest for the partner. If the drop is not too large the couple recovers to the positive equilibrium after the disturbance has ceased (see trajectory starting from point 1). But if the disturbance has brought the system into the other basin of attraction (see point 2), the couple will tend inevitably toward point E^-, characterized by pronounced and reciprocal antagonism. This is why such kind of couples have been called fragile. In conclusion, robust couples are capable of absorbing disturbances of any amplitude and sign, in the sense that they recover to a high quality romantic relationship after the disturbance has ceased. On the contrary, individuals forming a fragile couple can become permanently antagonist after a heavy disturbance and never recover to their original high quality mode of behavior.

The following property specifies when a couple is robust or fragile.

Property 2'
If the reactions I_1^* and I_2^* to the partner's appeal are sufficiently high, the couple is robust.

This property is rather intuitive: it simply states that very attractive individuals find their way to reconciliate.

4.2 Non secure - non synergic couples

We now consider couples composed of non secure and non synergic individuals. Thus, the return R_i and instinct I_i still depend only upon x_j and A_j, respectively. The resulting model can not have limit cycles (the proof immediately follows from Bendixon's criterion). This means that also these couples are characterized by steady romantic relationships.

4.3 Secure - synergic couples

One consequence of the results reported in the two preceding subsections is that couples composed of non synergic individuals, cannot have cyclic behavior. It can be also shown (see Gragnani et al.[15]), that the same holds for couples composed of secure individuals. Therefore *secure individuals cannot have cyclic love dynamics even if they are synergic.*

4.4 Non secure - synergic couples

In order to identify a case of cyclic dynamics we now consider couples composed of a secure and synergic individual and a non secure and non synergic individual. Thus, the model is

$$\dot{x}_1 = -\alpha_1 x_1 + R_1^*(x_2) + (1 + S_1^I(x_1))I_1^*(A_2)$$
$$\dot{x}_2 = -\alpha_2 x_2 + R_2^*(x_1) + I_2^*(A_1)$$

$$(4)$$

where R_1^*, R_2^*, I_1^* and I_2^* are shown in Fig. 1 (graphs (*ii*)), and S_1^I is like in Fig. 2.

In order to prove that model (4) can have cyclic dynamics, a numerical but rather systematic analysis of its local and global bifurcations (see Guckenheimer and Holmes[23]) has been performed in Gragnani *et al.*[15]

Figure 4 shows the bifurcation curves of model (4) in the two-dimensional space of synergism coefficient and maximum return of the first individual (a secure and synergic individual). The bifurcation curves divide the parameter space into five different regions. Each region is characterized by a different dynamic behavior, identified as 1,2, ..., 5 and described with a sketch of the corresponding state portrait. In regions 1, 2, 3 the attractor is unique (an equilibrium in regions 1 and 2 and a limit cycle in region 3), while in regions 4 and 5 there are alternative attractors. Without entering into the details, the diagram confirms our expectations: there are no cycles if the synergism coefficient is low but there are cycles if individual 1 is highly sensitive (high reaction to partner's love and high synergism coefficient). In conclusion, *a couple composed of a secure and synergic individual and a non secure and non synergic individual can have cyclic dynamics.*

The bifurcation diagram of Fig. 4 also shows that aging has a stabilizing effect. Indeed, it is generally believed that individual appeal, as well as synergism and reactions to partner's love and appeal, slowly deteriorate with aging. Thus, couples with cyclic love dynamics (regions 3 and 4 of Fig. 4) can slowly vary during their life and finally become stationary (regions 1, 2 and 5). This fact explains why cyclic romantic relationships characterized by relevant ups and downs tend to become more and more steady as life goes on.

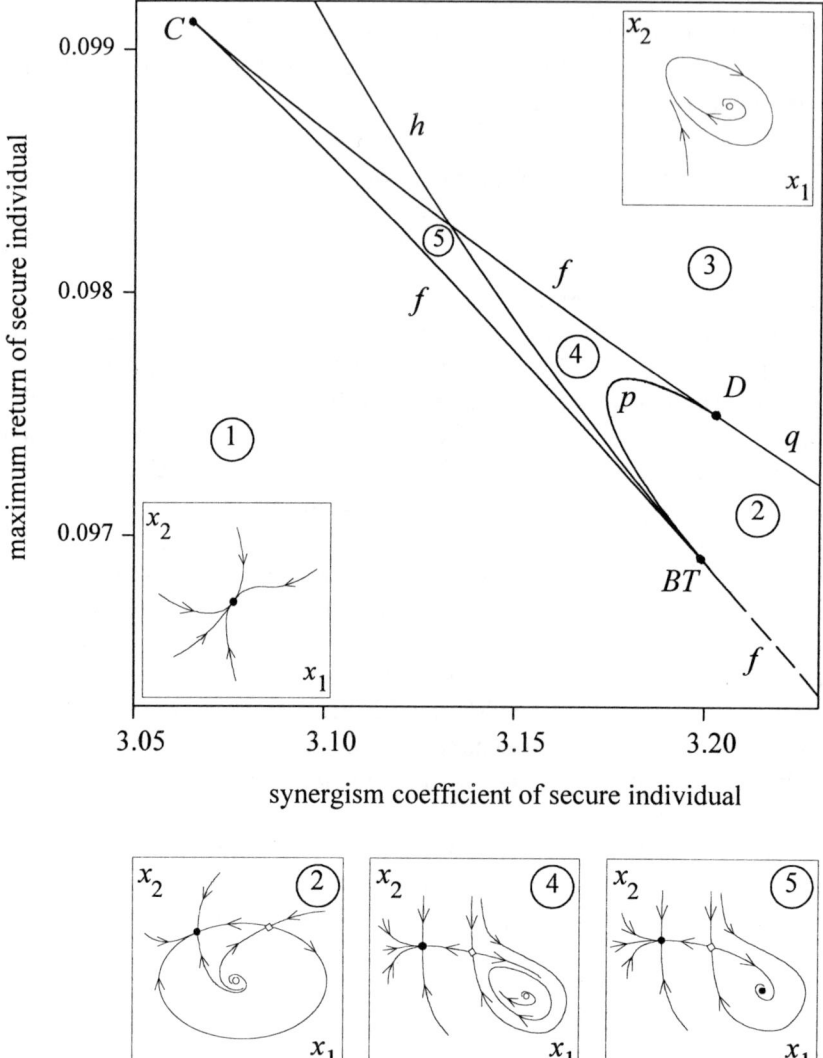

Figure 4 : Bifurcation portrait of model (4) in the two-dimensional parameter space of synergism coefficient and maximum return of the secure and non synergic individual.

5. A couple oriented study: the case of Laura and Petrarch

This section is devoted to the presentation of a model that pretends to describe the dynamics of love between two specific persons by accounting for their personalities. The model makes reference to a very special and famous case: the dynamics of love between Petrarch, an Italian poet of the 14-*th* century, and Laura, a beautiful and married lady. Their love story developed over 21 years and has been described in the *Canzoniere*, a collection of 366 poems addressed by the poet to his platonic mistress. The main traits of Petrarch's and Laura's personalities can be identified by analyzing the *Canzoniere*: he is very sensitive and transforms emotions into poetic inspiration; she protects her marriage by reacting negatively when he becomes more demanding and puts pressure on her, but at the same time, following her genuine Catholic ethic, she arrives at the point of overcoming her antagonism by strong feelings of pity. These traits can be encapsulated in a model composed of differential equations where the variables are the emotions of the two individuals. In particular, Laura can be described by a single variable, L, representing her love for the poet. Positive and high values of L mean warm friendship, while negative values should be associated with coldness and antagonism. The personality of Petrarch is more complex and its description requires two variables: P, love for Laura, and Z, poetic inspiration. High values of P indicate ecstatic love, while negative values stand for despair.

The model is an extension of the minimal model discussed in Sect. 2 and is described by the equations

$$\frac{dL(t)}{dt} = -\alpha_1 L(t) + R_L(P(t)) + \beta_1 A_P$$
$$\frac{dP(t)}{dt} = -\alpha_2 P(t) + R_P(L(t)) + \beta_2 \frac{A_L}{1 + \delta Z(t)}$$
$$\frac{dZ(t)}{dt} = -\alpha_3 Z(t) + \beta_3 P(t)$$

Note that the response of Petrarch to the appeal of Laura depends also upon his inspiration Z. This takes into account the fact that artistic inspiration attenuates the role of the most basic instincts. The third equation says that the love of Petrarch sustains his inspiration which, otherwise, would exponentially decay.

Petrarch's reaction function $R_P(L)$ is assumed to be linear $(\beta_2 L)$. On the contrary, Laura's reaction function $R_L(P)$ can be assumed to be linear only for low values of P, thus interpreting the natural inclination of a beautiful high-society lady to stimulate harmless flirtations. But Laura never goes too far beyond gestures of pure courtesy: she smiles and glances, but when Petrarch becomes more demanding and puts pressure on her, even indirectly when his poems are sung in public, she reacts very promptly and rebuffs him. This suggests the use of a reaction function $R_L(P)$ which, for $P > 0$, first increases and then decreases. But also for negative values of P the behavior of Laura is nonlinear. In fact, when $P \ll 0$, i.e., when the poet despairs, Laura feels very sorry for him. Following her genuine Catholic ethic she arrives at the point of overcoming her antagonism by strong feelings of pity, thus reversing her reaction to the passion of the poet.

The analysis of the model (see Rinaldi[16] for details) shows that its behavior tends toward a cyclic behavior. In other words, the peculiarities of the personalities of the two lovers, as they emerge from the poems of the *Canzoniere*, inevitably generate a never ending story of recurrent periods of ecstasy and despair. This is indeed what happened, as empirically ascertained by Jones[2] through a detailed stylistic and linguistic analysis of the dated poems (see Fig. 5). It is interesting to note, however, that with the modelling approach the existence of the emotional cycle is fully understood and proved to be inevitable, while with the empirical approach it is only discovered.

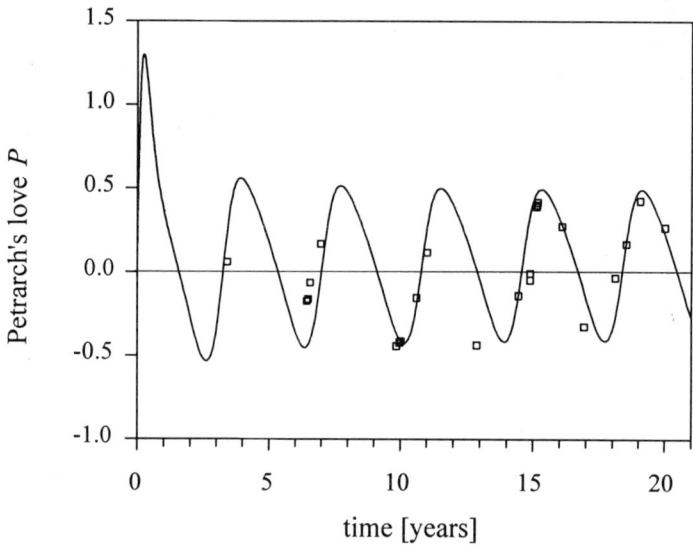

Figure 5 : Time evolution of the love of Petrarch predicted by the model and estimates (□) given by Jones[2].

6. Conclusions

Dynamics of love between two persons has been investigated in this paper within the framework of a minimal model composed of two differential equations, one for each individual. Three mechanisms of love growth and decay have been taken into account: the forgetting process, the pleasure of being loved and the reaction to partner's appeal. This has been done by introducing two functions, called return and instinct functions, which differ in the cases of secure and non secure individuals. The fact, here called synergism, that a woman [man] might react more strongly when she [he] is in love, has also been modelled. As a result, individuals are secure or non secure, and synergic or non synergic, for a total of four different classes. The review of the models studied up to now, has shown that couples composed of non synergic individuals as well as couples composed of secure individuals can not have cyclic dynamics. By contrast, couples composed of a secure and synergic individual and a non secure and non synergic individual can have cyclic dynamics. In other words, the presence of synergism and insecurity within the couple is the reason for tempestuous romantic relationships.

As for any minimal model, the extensions one could propose are innumerable. Aging, learning and adaptation processes could be taken into account allowing for some behavioral parameters to slowly vary in time, in accordance with the most recent developments of attachment theory. Men and women could be distinguished by using two structurally different state equations. The dimension of the model could also be enlarged in order to consider individuals with more complex personalities or the dynamics of love in larger groups of individuals. Undoubtedly, all these problems deserve further attention.

References

1. R.J. Sternberg and M.L. Baines in *Psychology of Love* (Yale U. P., 1988).
2. F.J. Jones in *The Structure of Petrarch's Canzoniere* (Brewer, Cambridge, 1995).
3. S.H. Strogatz, *Math. Mag.*, **61**, 35 (1988).
4. M.J. Radzicki, *Sys. Dyn. Rev.*, **9**, 79 (1993).
5. S.H. Strogatz in *Nonlinear Dynamics and Chaos with Applications to Physics, Biology, Chemistry and Engineering* (Addison-Wesley, Reading, MA, 1994).
6. S. Rinaldi, *Appl. Math. Comp.*, to appear (1997).

7. A. Berman and R.J. Plemmons in *Nonnegative Matrices in the Mathematical Sciences* (Academic Press, New York, 1979).

8. D.G. Luenberger in *Introduction to Dynamic Systems* (Wiley, New York, 1979).

9. A. Graham in *Nonnegative Matrices and Applicable Topics in Linear Algebra* (Ellis Horwood Limited, Chichester, 1987).

10. A. Berman *et al.*, in *Non-negative Matrices in Dynamic Systems* (Wiley, New York, 1989).

11. S. Rinaldi and L. Farina in *Positive Linear Systems: Theory and Applications* (Utet, Torino, 1995).

12. S. Rinaldi and A. Gragnani, *Nonlin. Dyn. Psych. Life Sci.*, submitted (1997).

13. K. Bartholomew and L.M. Horowitz, *J. Pers. Soc. Psych.*, **61**, 226 (1991).

14. D.W. Griffin and K. Bartholomew, *J. Pers. Soc. Psych.*, **67**, 430 (1994).

15. A. Gragnani *et al.*, *Int. J. Bif. Chaos*, to appear (1997).

16. S. Rinaldi, *SIAM J. Appl. Math.*, to appear (1997).

17. A. Feingold, *J. Pers. Soc. Psych.* **59**, 981 (1990).

18. S. Sprecher *et al.*, *J. Pers. Soc. Psych.*, **66**, 1074 (1994).

19. R.R. Kobak and C. Hazan, *J. Pers. Soc. Psych.*, **60**, 861 (1991).

20. E. Scharfe and K. Bartholomew, *Pers. Rel.*, **1**, 23 (1994).

21. T.L. Fuller and F.D. Fincham, *Pers. Rel.*, **2**, 17 (1995).

22. J.A. Simpson *et al.*, *J. Pers. Soc. Psych.*, **59**, 1192 (1990).

23. J. Guckenheimer and P. Holmes in *Nonlinear Oscillations, Dynamical Systems and Bifurcations of Vector Fields* (Springer Verlag, New York, Heidelberg, Berlin, Tokyo, 1983).

FRACTAL DYNAMICS OF HEARTBEAT INTERVAL FLUCTUATIONS IN HEALTH AND DISEASE

M. MEYER, C. MARCONI, A. RAHMEL, B. GRASSI, G. FERRETTI, J.E. SKINNER, P. CERRETELLI

Département de Physiologie, CMU, Genève, Switzerland
Istituto di Tecnologie Biomediche Avanzate, CNR, Milano, Italy
Department of Physiology, Max Planck Institute for Experimental Medicine,
Göttingen, Germany
Totts Gap Laboratories, Bangor, USA

The dynamics of heartbeat interval time series were studied by a modified random walk analysis recently introduced as *Detrended Fluctuation Analysis*. In this analysis, the intrinsic fractal long-range power-law correlation properties of beat-to-beat fluctuations generated by the dynamical system (i.e. cardiac rhythm generator), after decomposition from extrinsic uncorrelated sources, can be quantified by the scaling exponent which, in healthy subjects, is about 1.0. The finding of a scaling coefficient of 1.0, indicating scale-invariant long-range power-law correlations ($1/f$ noise) of heartbeat fluctuations, would reflect a genuinely self-similar fractal process that typically generates fluctuations on a wide range of time scales. Lack of a characteristic time scale suggests that the neuroautonomic system underlying the control of heart rate dynamics helps prevent excessive mode-locking (error tolerance) that would restrict its functional responsiveness (plasticity) to environmental stimuli. The $1/f$ dynamics of heartbeat interval fluctuations are unaffected by exposure to chronic hypoxia suggesting that the neuroautonomic cardiac control system is *preadapted* to hypoxia. Functional (hypothermia, cardiac disease) and/or structural (cardiac transplantation, early cardiac development) inactivation of neuroautonomic control is associated with the breakdown or absence of fractal complexity reflected by anticorrelated random walk-like dynamics, indicating that in these conditions the heart is *unadapted* to its environment.

1 Introduction

The cardiac rhythm results from the spontaneous activity of specific cardiac cells located in a small area of the right atrium (sino-atrial node). The activity is modulated by the activity of the sympathetic and parasympathetic components of the autonomic nervous system. The precise mechanisms of interaction which may further be modulated by other cardiac and extracardiac factors have not been established but offer the possibility of interaction of several negative or mixed feedback systems with delays that have been demonstrated to produce periodic, quasiperiodic or even chaotic dynamics.[4]

1.1 Fractal Scaling of Heart Rate

The long-term variability of heart rate observed over a wide range of time scales with scale-invariant power-law characteristics ($1/f$ noise) has recently been associated with the fractal scaling behavior and long-range correlation properties typically exhibited by dynamical systems near a critical point of a phase transition, i.e. far away from equilibrium.[1,13,15,20] The fractal scaling exponent (α) of normal 24 hr heartbeat interval time series (about 10^4 beats) calculated by a novel algorithm (*Detrended Fluctuation Analysis, DFA*) was ~ 1.0, indicating the presence of non-trivial long-range correlations (memory-like effects) while patients with congestive heart failure showed a significant deviation of the long-range correlation exponent from normal (~ 1.25). The physiological significance of the long-range power-law or $1/f$ behavior of normal cardiac interbeat time series has not been established. The physiological advantage of a non-equilibrium dynamic system operating over a wide range of time scales, unlike a classic homeostatic process, may be seen in its capacity to respond to external or environmental stimuli (*plasticity*) while preserving a relative insensitivity to errors (*error tolerance*).

1.2 Hypothesis

In pursuit of this concept, we hypothesized that the normal intact heart, in terms of the intrinsic mechanisms controlling heart rate and its fluctuations exhibits a high degree of fractal complexity and is *preadapted* to external stimuli such as hypoxia encountered during exposure to a hypobaric environment, i.e. high altitude. On the contrary, functional (e.g. hypothermia, cardiac disease) or structural inactivation (cardiac transplantation, early cardiac development) of neuroautonomic cardiac control is expected to be associated with the breakdown or absence of fractal complexity, indicated by the emergence or prevalence of no (or trivial) correlations ("anti-correlations").

1.3 Objectives

The major objectives of the present studies were: *1*) to quantify the scaling exponent and correlation properties of stationary and non-stationary heartbeat interval time series over long-term (24 hr) and short-term (40 min) epochs; *2*) to assess the stability or otherwise adaptive changes that would result from changes in the physiological environment, e.g. chronic exposure to a hypoxic environment (high altitude), long-term hypothermia, early cardiac development, or in disease, e.g. congestive heart failure and chronic cardiac denervation after transplantation; *3*) to evaluate the mechanisms and the physiological significance underlying the fractal complexity of heart rate dynamics.

2 Methods

2.1 Subjects

Quantitative assessment of cardiac dynamics is based on heartbeat interval time series derived from ECG monitoring. The measurements were performed in normal healthy subjects, in patients with congestive heart failure or after cardiac transplantation, in experimental animals and in the developing chick embryo. A group of normal subjects were members of a medical high altitude expedition sojourning for 34 days at the Italian EvK2-CNR high altitude Pyramid Laboratory (Sagarmatha National Park, Khumbu Valley, Nepal, 5050 m - 16568 ft.; $P_b \sim 420$ mmHg, $P_{IO_2} \sim 79$ mmHg) located in the vicinity of the Everest Base Camp.

2.2 ECG Acquisition

Long-term (24 hr) electrocardiographic recordings were obtained by ambulatory Holter recorders provided with RAM-memory and advanced data compression technology. The 24 hr dual-channel digitized ECG data (sampling rate 500 Hz per channel) were automatically processed and annotated for established clinical electrocardiographic criteria. Short-term recordings (40 min) were obtained at rest or during an exercise protocol using a dual-channel battery-powered miniature ECG recorder-amplifier system (sampling rate 1200 Hz per channel) interfaced to a PC by a fiber-optical link.The heartbeat interval time series were obtained automatically by beat recognition and an adaptive R-wave threshold detection algorithm.

2.3 Fractal Scaling Exponent

A novel modified random walk analysis, termed *Detrended Fluctuation Analysis (DFA)*, has recently been introduced by Peng and coworkers to study the fluctuations and order of DNA sequences [3,14] and subsequently applied to human gait [6,7] and heart rate times series.[13,15] The analysis is based on the observation that heart rate fluctuates considerably even in the absence of fluctuating external stimuli rather than relaxes to a homeostatic steady state. The fluctuations over minutes resemble those over hours or days, i.e. the beat-to-beat fluctuations of heart rate appear to be *self-similar* on different time scales. The concept of the analysis is based on the statistical self-similarity in a signal. A true one-dimensional signal (the signal is a function of time) may be characterized as self-similar or fractal if its subsets can be rescaled to resemble statistically the original sequence itself. A quantitative measure of this scaling procedure is defined by the scaling exponent or self-similarity parameter. A stationary time series with long-range temporal correlations can be integrated to form a self-similar process.[2,16] Hence, determination of the self-similarity scaling exponent yields the inherent long-range correlation properties of the original series.

Under free-running conditions the heartbeat interval time series is typically highly non-stationary as reflected by the local average varying with time. The fluctuations in the heart rate pattern may represent: *1)* uncorrelated (white) noise superimposed on a basically regular process; *2)* short-range correlations such that the current value is influenced only by its most recent predecessors but otherwise random fluctuations over the long term; *3)* long-range correlations generated by the intrinsic complex non-linear dynamics of the system itself; *4)* trivial changes of physiological or environmental conditions (unsteady state). A key issue to the analysis of non-stationary physiological time series data is that fluctuations driven by uncorrelated stimuli can be interpreted as systematic "shifts" or "trends" related to the frequency of the stimuli and distinguished from the intrinsic multi-component correlation properties of the dynamics.

In the DFA analysis the integrated time series ($y(k)$, length N) is self-similar if the fluctuations at different observation windows ($F(n)$) scale as a power-law with the window size (n), i.e. the number of intervals in the window of observation. Non-stationarities are treated by removing the least-squares linear regression trend in each window. The root-mean-square fluctuation ($F(n)$) is calculated for all window sizes:

$$F(n) = \sqrt{\frac{1}{N} \sum_{k=1}^{N} [y(k) - y_n(k)]^2} \qquad (1)$$

where $y_n(k)$ denotes the local trend in each window. A linear relationship in the log $F(n)$ vs. log n graph indicates that $F(n) \sim n^{\alpha}$, where α (slope of the log $F(n)$ vs. log n relationship) is the scaling exponent (or self-similarity parameter).

2.4   Correlation Properties of Time Series

The scaling exponent is related to different correlation properties of the time series: *1)* for white noise which is composed of a sequence of independent random variables, $\alpha = 0.5$; *2)* for Brownian noise (classical random walk) which results

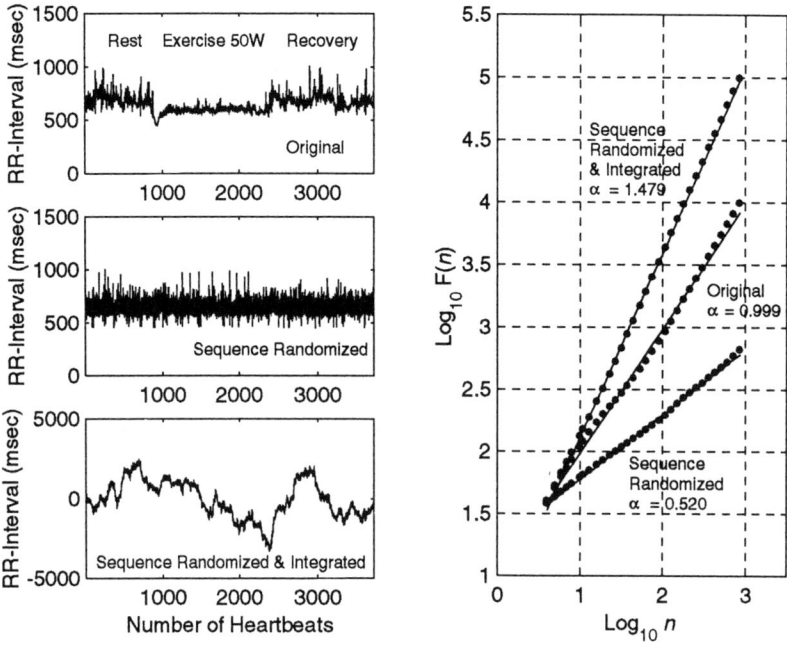

Figure 1: Correlation properties of heart beat interval time series. Original experimental time series (*upper left*); Sequence-randomized surrogate data set (*middle left*); Sequence-randomized and integrated data set (*lower left*). Scaling coefficients of time series (*right*).

from the integration of white noise where the individual steps are random and uncorrelated, $\alpha = 1.5$; *3*) persistent long-range correlations exist for $0.5 < \alpha \leq 1.0$; *4*) for $1/f$ noise which represents the boundary between stationarity ($\alpha < 1.0$) and non-stationarity ($\alpha > 1.0$), $\alpha = 1.0$. The finding of long-range correlations is indicative of self-similarity, scale-invariance and a fractal pattern. Long-range correlations indicate that, on average, the fluctuations on one time scale are self-similar to those on other time scales.

A typical example of a non-stationary heartbeat interval time series is shown in Fig. 1. The original time series derived from 40 min continuous ECG recording (*upper left*) comprises three subepochs: *1*) resting conditions; *2*) square-wave cyclic exercise (50 W) indicated by an almost instantaneous decrease of heartbeat interval durations (increase of heart rate); *3*) recovery from exercise. The log F(n) vs. log n plot (*right*) of the integrated and detrended original time series yields a straight line with slope $\alpha \sim 1.0$. Thus, the heartbeat interval fluctuations scale as $F(n) \sim n^{1.0}$, indicating persistent non-trivial long-range correlations that are not the consequence of summation over random variables or artifacts of non-stationarity due to different levels of physical activities. The sequential correlated ordering in the original time series may be eliminated by constructing a sequence-randomized surrogate data set (*middle left*) with identical heartbeat interval distribution (i.e. same mean, SD, and higher moments) which demonstrates uncorrelated white noise ($\alpha \sim 0.5$). Integration of the sequence-randomized surrogate set yields Brownian noise (*lower left*) with α approaching 1.5. If $0.5 < \alpha \leq 1.0$, the time series is correlated in time such that longer (shorter) heartbeat intervals are likely to be close to each other. If $1.0 < \alpha \leq 1.5$, the time series is also correlated but the heartbeat interval sequence is more likely to alternate between long and short intervals ("anti-correlation"). Both white noise and its integral, Brownian noise, are examples of true random processes with no (or trivial) dependence on their past history. The heartbeat interval time series produces a contour reminiscent of irregular "landscapes". The moderately rough landscape of $1/f$ fluctuations ($\alpha \sim 1.0$) may be interpreted as a "compromise" between the very rough landscape of white noise ($\alpha \sim 0.5$) and the very smooth landscape of Brownian noise ($\alpha \sim 1.5$).

2.5 Experimental and Simulated Time Series

The DFA analysis was applied to the experimental heartbeat interval series of the full-length 24 hr epochs. Typically, the data length ranged from 90000 to 135000 data points as a result of varying daily activities. In addition, 4 hr subsets from 1:00 am (pm) through 5:00 am (pm) were selected from the original time series to determine the scaling properties of night-time and day-time subepochs that were 12 hrs apart and differed in the subject's physical activity. The data length of these segments was 13000 to 17000 data points. The data length of the short-term (40 min) recordings was about 4500 points.

In order to assess the effects of data length and of noise added to the signal on the accuracy and precision of estimates of the scaling coefficient α, genuine fractal time series with known correlation coefficients were generated with number of points ranging from $N = 1024$ (2^{10}) to $N = 131072$ (2^{17}) and α ranging from 0.5 to 1.5.[11,21] Ten realizations were obtained for each N and α. For the noise analysis, Gaussian random noise was added to the genuine series with levels ranging from 5 to 500% of the standard deviation of the original signal. For statistical comparison, fitting of the scaling exponent α was restricted to a least-squares fit of $\log F(n)$ vs. $\log n$ for $16 \leq n \leq N/2$, thus eliminating very large asymptotic time scales. For finite-length data sets the error of $F(n)$ has been demonstrated to increase with n and scaling exponents calculated over a larger range of n have less accuracy.[12]

3 Results

3.1 Experimental Time Series

Figure 2 (*upper*) shows the DFA analysis of a representative 24 hr heartbeat interval time series ($N = 102652$ points) demonstrating power-law scaling over more than three decades with $\alpha \sim 1$, i.e. $1/f$ noise. For comparison, a genuine stationary sequence (same data length, mean and dynamic range) with $\alpha = 1.0$ was generated (Fig. 2, *lower*). Visual comparison of the fluctuations of the experimental and genuine time series remains ambiguous; nevertheless, the DFA analysis reveals that the correlation properties of the two series are close to identical.

Figure 3 shows the $\log F(n)$ vs. $\log n$ plots of serial 24 hr heartbeat interval time series from an individual subject during sojourn at high altitude. The DFA analysis demonstrates that the α exponents (slopes of the regression lines for $16 \leq n \leq N/2$) were close to 1.0 (overall mean $\pm$ SD, 1.04 ± 0.2, range 0.99 to 1.07) and were unaffected by the subject's exposure to a hypobaric (hypoxic) environment.

A survey of the results of the DFA analysis of serial 24 hr heartbeat interval time series and its 4 hr subsets of a healthy subject along with the altitude profile and locations, covering sea level, ascent, sojourn at 5050 m for 34 days, and return from altitude, is shown in Fig. 4. The scaling exponent of the 24 hr epochs varies from about 0.99 to 1.08 but is unaffected by the subject being exposed to hypoxia. No differences are apparent when the data from the initial stages of exposure to hypoxia are compared with those of the later stages. The overall mean value $\pm$ SD of α (1.04 ± 0.03) is statistically not different from 1.0, suggesting that heart rate fluctuations exhibit $1/f$ dynamics. The overall mean values $\pm$ SD of the 4 hr subepochs were 0.97 ± 0.06 (day-time) and 0.88 ± 0.05 (night-time), respectively. These findings would suggest an apparent inconsistency of the DFA analysis which may be resolved by evidence that the 24 hr time series exhibits *composite* rather than *unique* fractal properties (*see Sec. 4.4*). Of relevance here is the observation that both day-time and night-time sequences do not show any systematic changes in response to high altitude. No systematic directional changes of α with exposure to hypoxia were observed in any of the 9 subjects. The group mean average $\pm$ SD was 0.99 ± 0.04, the group mean day-night difference of the 4 hr subepochs 0.10 ± 0.05.

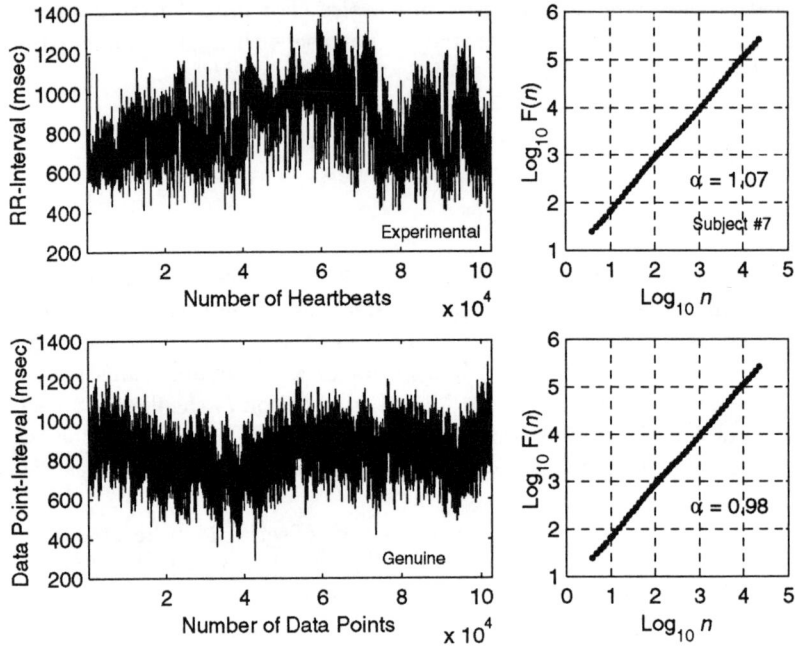

Figure 2: Detrended fluctuation analysis of 24 hr heartbeat interval time series (*upper*) and genuine long-range power-law correlated fractal time series (*lower*). Scaling coefficients of time series (*right*).

3.2 Scaling Properties in Health and Disease

The scaling properties of heartbeat interval time series in health and disease are compiled in Tab. 1. The following features are important.

Normal Subjects

In normal adult subjects the scaling coefficients designate the prevalence of $1/f$ dynamics ($\alpha \sim 1.0$) no matter whether the subject is in steady or unsteady state conditions. In contrast, the heart rate dynamics in children demonstrated long-range power-law correlations with $\alpha = 0.75$ but the $1/f$ dynamics of adults with $\alpha = 1.0$ would not be achieved until maturation.

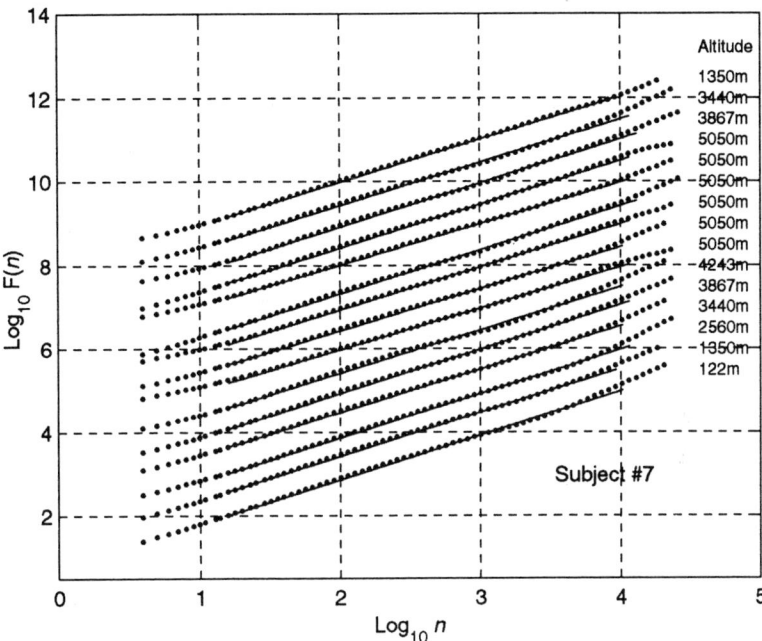

Figure 3: Plot of log $F(n)$ vs. log n of 24 hr heartbeat interval time series of a subject sojourning at high altitude. The regression slopes for $16 \leq n \leq N/2$ are shown by solid lines. The plots are shifted vertically for the purpose of display.

Chronic Hypoxia

Physiological systems, as a strategy of adaptation, are known to undergo functional and structural changes in response to external challenge, e.g. chronic hypoxia. Physiological adaptation, defined here as a change of a physiological variable which reduces the strain imposed by stressful components of the environment, is known to be time-dependent and would be expected to elicit differential responses that rely on instantaneous strategies in the acute phase and on acclimatization in the chronic phase. In contrast, the mechanisms underlying the control of heartbeat interval dynamics demonstrate lack of any adaptation to hypoxia.

114

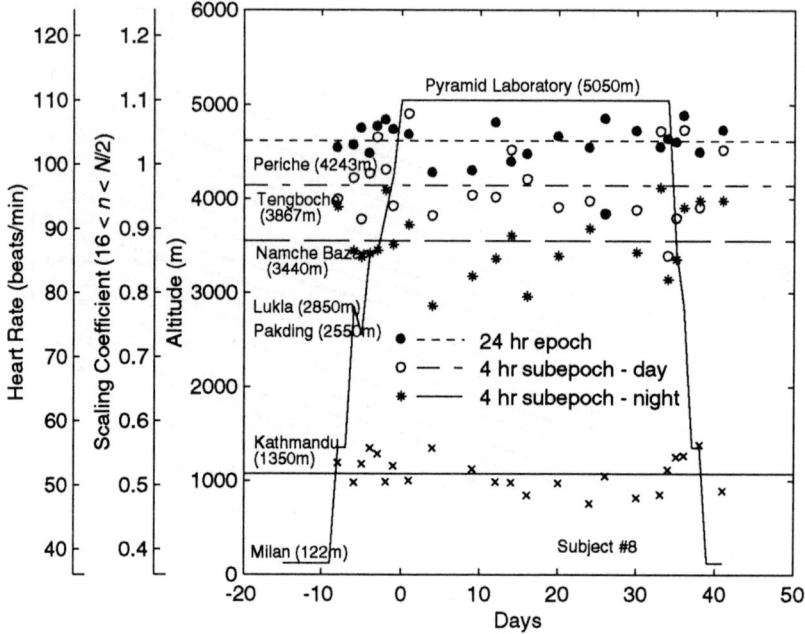

Figure 4: Scaling coefficients of 24 hr, 4 hr day-time and 4 hr night-time heartbeat interval time series of a subject during acclimatization to high altitude (5050m). The locations and altitudes of the measurements are indicated. The arrival at the Pyramid Laboratory is arbitrarily assigned as day 0. Resting heart rate was calculated from 4 hr night-time subepochs. Horizontal lines are means.

Hypothermia and Embryonic Heart

Functional inactivation of autonomic cardiac control by prolonged hypothermia (28°C) or lack of functional or structural evolution of the autonomic nervous system during the early stages of development causes the cardiac rhythm to follow uncorrelated random walk-like dynamics with $\alpha \sim 1.5$.

Heart Failure

In a group of congestive heart failure patients the group averaged a exponent was 1.24 ± 0.16, in agreement with previous results by Peng and coworkers.[13,15] It is important to emphasize that for $1.0 < \alpha \leq 1.5$ correlations exist but cease to be of power-law type.

Table 1. Scaling Coefficient of Heartbeat Fluctuations in Health and Disease.

n	α	Remarks
	Normal Adults	
17	1.01 ± 0.11	unsteady state
14	0.99 ± 0.07	steady state, supine
9	0.99 ± 0.04	unsteady state, 24 hr epochs,
	$0\text{-}96 \pm 0.05$	4 hr day-time subepochs
	0.86 ± 0.04	4 hr night-time subepochs
		chronic high altitude hypoxia
	Normal Children	
8	0.75 ± 0.08	steady state, supine
	Chick Embryo	
28	1.50 ± 0.14	2 - 8 days of a 21 - day
		incubation period
	Pigs	
4	1.51 ± 0.05	Hypothermia, 28°C, 48 - 72 hrs
	CHF	
20	1.24 ± 0.16	unsteady state
	HTR - Paris	
28	1.54 ± 0.18	unsteady state
	HTR - Bergamo	
13	1.48 ± 0.11	< 2 yrs after transplantation
17	1.02 ± 0.16	> 2 yrs after transplantation
		steady state, supine,
		Adults and Children

Values are means. n, number of measurements. All data were obtained by 40 min ECG recording sessions except stated otherwise. In unsteady state conditions the subjects underwent an exercise protocol. CHF, patients with congestive heart failure; HTR - Paris, heart transplant recipients operated at La Pitie-Salpetrière Hospital, Paris, 1981 - 1990; HTR - Bergamo, heart transplant recipients operated at Riuniti Hospital, Bergamo, 1987 - 1995.

Heart Transplantation

A straightforward approach as to the significance of autonomic cardiac control for the fractal dynamics of heart rate is facilitated by studies in recipients of a cardiac transplant. The α exponent in heart transplant recipients (HTR) was ~ 1.5 early (< 2 yrs) after transplantation, indicating that the denervated heart was operating persistently in the limiting-state random walk regime.

The process underlying the heart rate dynamics of the denervated heart has no memory of the past (*see Sec. 4.6*). For the later stages after transplantation (> 2 yrs) the α exponents approached ~ 1.0 suggesting that the dynamics had moved away from the random walk regime as a result of recovery of neuroautonomic cardiac control, but in children would not attain normal dynamics in long-term survivors of cardiac transplantation.

3.3 Neuroautonomic Cardiac Control and Fractal Dynamics

The functional evidence for the recurrence of cardiac control late after transplantation addresses the issue of its structural representation. In the intact heart control of heart rate is governed by the *extrinsic* nervous system, i.e. the branches of the central nervous system, and the *intrinsic* cardiac nervous system located at the base of the heart containing efferent post-ganglionic sympathetic and parasympathetic neurons, local circuit and afferent neurons. After surgical decentralization the intrinsic canine cardiac neurons have been demonstrated to retain some capacity to modulate the heart.[10,19] Thus, functional recurrence of heart rate control after cardiac transplantation may be due to reinnervation of the transplanted allograft by the extrinsic nervous system but preliminary evidence suggests that the post-transplant recovery of heart rate fluctuations may be due to reorganization or adaptation of viable intrinsic cardiac neurons.[9] Interestingly, the HTR-Paris group, unlike the HTR-Bergamo group, did not demonstrate any recurrence of fractal dynamics up to 8 yrs after transplantation which may be attributed to differences in the induction of immunosuppressive therapy and the effects on the viability of intrinsic cardiac neurons rather than to differences in cardiosurgical procedures. It may be noted that the transplanted heart which, in terms of its heart rate control, is deprived of its fractal complexity may well serve its primary function, i.e. providing for a sufficient cardiac output, by compensatory mechanisms of myocardial contractility and denervation *per se* is unlikely to account for the impaired exercise capacity of cardiac transplant patients.[5,8]

Based on these findings we hypothesize that the intrinsic long-range correlations of the dynamics of physiologic interbeat fluctuations are generated by the autonomic nervous system projecting onto the cardiac rhythm generator.

3.4 Genuine Fractal Time Series With Known Correlation Properties

Accuracy, Precision and Data Length

Artificial stationary fractal time series with known α from 0.5 to 1.5 and lengths $N = 131072$ (2^{17}) points were generated to assess the accuracy (defined by the

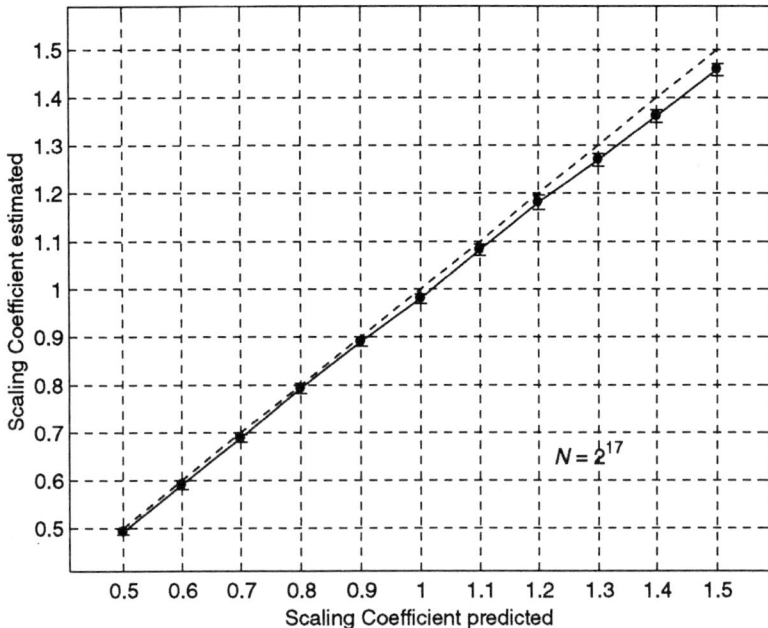

Figure 5: Estimation of the scaling coefficient α (*ordinate*) for genuine stationary fractal time series with known correlation properties (*abscissa*). The results $\pm$ SD are shown for 10 realizations at each α for $0.5 \leq \alpha \leq 1.5$ and data length of 131072 (2^{17}) points. Dotted line is line of identity.

SD of the estimate) and precision (defined as absence of bias) of the DFA analysis. Ten realizations were obtained for each α and N. Shorter series from $N = 1024$ (2^{10}) to $N = 65536$ (2^{16}) were selected as part of the original series.

Figure 5 shows the results (means $\pm$ SD for 10 realizations each) of estimation of the scaling coefficients for time series with $N = 2^{17}$ points for α ranging from 0.5 to 1.5. The absolute SD of the α estimates is about 0.0065. Hence, the accuracy of the estimates is about 0.4%, 0.6% and 1.3% for $\alpha = 1.5$, $\alpha = 1.0$ and $\alpha = 0.5$, respectively. The resolution, $\Delta\alpha$, defined as 3 times the SD, for time series of length $N = 2^{17}$ (corresponding to the length of the experimental 24 hr heartbeat interval series) and $\alpha \sim 1.0$ is about 0.02. However, there is a small systematic error or bias in the α estimates as the slope of the $\alpha_{estimated}$ vs. $\alpha_{predicted}$ relationship is slightly less than unity (0.97). Thus, the estimates of α are lower by about 3% of the correct value over the range $0.5 \leq \alpha \leq 1.5$.

For shorter artificial time series with α values from 0.5 to 1.5 and lengths N from 2^{10} to 2^{17} the results are similar. The SD is larger for short series ($\sim$ 0.06 for $N = 2^{10}$) and decreases as a power-law function of N ($\sim$ 0.01 for $N = 2^{17}$) but the bias is about constant at $\leq 3\%$ over the range of $2^{10} \leq N \leq 2^{17}$. Scaling coefficients derived from shorter experimental data sets may thus be compared with those of longer series but the results from shorter series are less reliable. The source of the small bias in the estimates of α is unclear and may be circumstantial due to inaccuracies of the method of signal generation, of the DFA analysis, or both. Notwithstanding, the potential systematic error of the DFA analysis when applied to experimental data is small.

Effects of Noise

The effects of noise on the estimates of α (for $16 \leq n \leq N/2$) were studied by the addition of Gaussian white noise ($\alpha = 0.5$) with zero mean and varied SDs from 5 to 500% of that of the original fractal time series for $N = 2^{10}$ to $N = 2^{17}$ points. Added noise drives the effective correlation toward zero and the estimate of α toward 0.5 but the overall effects are relatively small. The effects are less pronounced for series with larger N; the estimates of α remain closer to the true values at higher noise levels for $N = 2^{17}$ than for $N = 2^{10}$ points. Fractal signals recorded with noise levels as large as 1.5 times the SD of the original signal appear to be safely distinguishable from true noise. In particular, for $\alpha = 1.0$ and for large N the estimates of α are well determined from a single realization and are only moderately degraded even with an addition of 100% Gaussian noise. It is important to note that long-range correlations with power-law characteristics cannot trivially be generated from fractal signals modulated by multiple time scales. For example, the superposition of white ($\alpha = 0.5$) and brown ($\alpha = 1.5$) noise would not produce long-range correlations or $1/f$ dynamics ($\alpha = 1.0$) no matter how the amplitudes of these components are varied.

In summary, the DFA analysis is quite robust in three points: *1)* proper estimates of the correlation properties of the time series may be obtained from series as short as 1024 points, *2)* the bias in the estimated scaling coefficient α is small ($\leq 3\%$), and, *3)* the method is not highly sensitive to noise.

4   Discussion

4.1   Scaling Properties of Heartbeat Interval Fluctuations

The analysis of scaling properties of long-term 24 hr heartbeat interval time series has revealed three important findings. *1)* The long-term fluctuations of heart rate exhibit long-range, scale-invariant, power-law characteristics (scaling exponent $\alpha \sim 1$, i.e. $1/f$ noise) reminiscent of dynamical systems far away from equilibrium.[1,20] 2) The intrinsic dynamics of the heartbeat interval fluctuations, i.e. $1/f$ behavior, is essentially unaffected by the physiologic challenge of chronic exposure to hypoxia. The results are consistent with the hypothesis that the intrinsic mechanisms generating fractal $1/f$ dynamics of heartbeat fluctuations are *preadapted* to chronic hypoxia, i.e. the intrinsic heartbeat dynamics do not undergo early readjustment or late acclimatization. 3) The analysis of the 4 hr day-time and 4 hr night-time subepochs selected as subsets of the full-length 24 hr time series has revealed an apparent discrepancy in that the scaling properties of the heartbeat interval series during night-time episodes were lower ($\alpha = 0.86 \pm 0.04$) compared to the day-time episodes ($\alpha = 0.96 \pm 0.05$) or the 24 hr epochs ($\alpha = 0.99 \pm 0.04$). These findings suggest that the fractal characteristics of heart rate in the course of a 24 hr period of observation were evolving with time and became local, i.e. the signal is what is generally referred to as a *multifractal*. However, the multi-component composite fractal nature of the time series would remain unnoticed by the DFA analysis when applied to the full-length 24 hr data base (*see Sec. 4.4*). 4) Break-down or absence of fractal long-range power-law correlations of the heartbeat interval time series is observed in non-physiological conditions (hypothermia), in cardiac disease (congestive heart disease, cardiac transplantation), or during the early stages of cardiac development. In these conditions the intrinsic cardiac dynamics may be considered as "anti-correlated" or *unadapted* to the environment.

4.2 Non-Stationarity and Fractal Dynamics

Conceptually, physiological time series such as heartbeat interval time series under free-running conditions may be visualized as "badly behaved" data originating from non-stationarities driven by uncorrelated extrinsic factors and/or manifesting the intrinsic non-linear correlated dynamics of a fractal process. Hence, any analysis of heartbeat interval time series is confronted with the problem of distinguishing a stationary fractal time series from a non-stationary process.

Formally, the signal generated by a fractal process with fixed components is non-stationary, i.e. it varies all over as a result of its "built-in" long-range power-law properties and its moments depend on time. A self-similar fractal process typically produces a time series with long-range correlations of the fluctuations that display the kind of "drift-like" appearance at all time scales. The statistical properties of a fractal process are characterized by an infinite mean or infinite variance and the mean of a measured property changes with the amount of data analyzed and would not converge to a well defined value. Hence, non-stationarity only indicates that the moments are not defined but would not imply that the process that generated the data was changing in time. The situation is further complicated by the fact that a fractal process may not be completely described by a unique scaling coefficient (or single fractal dimension) because it constitutes a composite of fractals with varying scaling properties (or different fractal dimensions).

On the other hand, the intrinsic dynamics of a complex non-linear system may be biased by extrinsic sources, i.e. non-steady state physiological or environmental conditions that could give rise to highly non-stationary behavior. Although these strain-related variations may be physiologically important, their correlation properties would be expected to be different (related to the stimulus) from the long-range correlations ("memory") generated by a dynamical system.

Typically, the 24 hr heartbeat time series of the present study is highly non-stationary with heart rates ranging from 40/min (night-time supine resting conditions) to about 200/min (day-time physical activity). A key feature of the detrended fluctuation analysis is that the extrinsic fluctuations from uncorrelated stimuli can be interpreted as "trends" and decomposed from the intrinsic dynamics of the system itself. The principle of calculating a measure of dispersion ($F(n)$) as function of window size (n) consequent upon subtracting the local trend in each window facilitates the detection of intrinsic long-range correlations embedded in a non-stationary time series.

In Fig. 6 the log $F(n)$ vs. log n plots are shown for an experimental time series ($\alpha = 1.032$) and 10 realizations of a genuine stationary fractal process with the same scaling exponent. The regression slopes ($16 \leq n \leq N/2$) of the genuine time series (mean $\pm$ SD, 1.036 ± 0.018, range 1.009 to 1.061) were not statistically different (t-test of regression slopes) from the experimental series, suggesting that the experimental series was stationary and generated by a fractal process with $1/f$ dynamics (strictly, this is not proof but strong evidence).

4.3 Detrending Fractal Time Series

Unlike conventional methods (e.g. power spectrum analysis and Hurst analysis) that presume stationarity of the data stream, the DFA technique does not rely on this assumption. Indeed, the persistence of long-range correlations after detrending would be indicative of a genuinely self-similar process. In order to rule out the possibility that the fluctuations in the experimental time

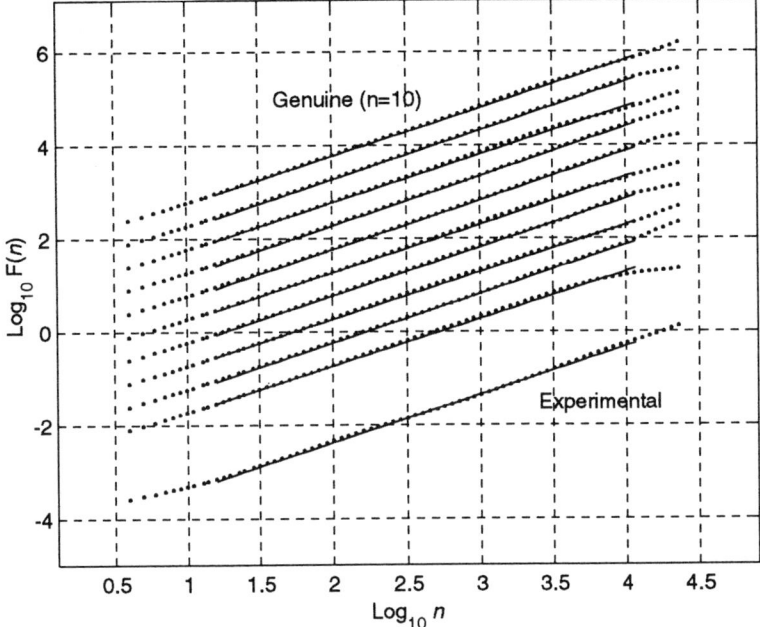

Figure 6: Plot of log $F(n)$ vs. log n for a typical experimental 24 hr heartbeat interval time series and 10 realizations of a genuine fractal time series with the same length and scaling coefficient. The regression slopes for $16 \leq n \leq N/2$ are shown by solid lines. The plots are shifted vertically for the purpose of display.

series and their long-range correlations were simply a trivial consequence of extrinsic non-stationarities, the data series were detrended prior to applying the DFA analysis. A typical example is shown in Fig. 7 (*left*). Detrending was performed by fitting the data by polynomials of order 1 to 5 and then by subtracting the fitted curve from the data. The procedure was performed at multiple time scales, i.e. the original time series was detrended for segments ranging from $2^{10} \leq N \leq 2^{17}$ data points. In the example of Fig. 7 (*left*) the estimated scaling coefficient α for the experimental time series ($N = 131072$ data points) prior to detrending was 1.04 and for segments varying from $2^{16} \geq N \geq 2^{10}$ points remained practically unchanged (range 0.99 to 1.05) but the SD of the estimates increased with shorter data sets. The lower and higher order detrending is expected to remove much of the slow varying "drifting" in the time series but has almost no effect on the estimated scaling coefficient

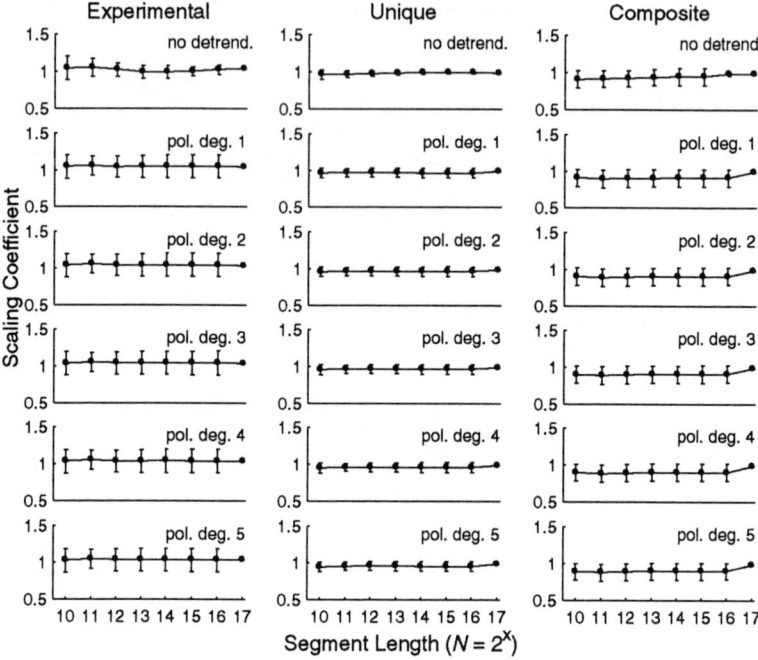

Figure 7: Effects of detrending of experimental (*left*) and fractal time series with unique (*middle*) or composite (*right*) fractal properties.

because the short-scale fluctuations would remain practically unchanged.

For comparison, the results of the detrending procedures are shown in Fig. 7 (*middle*) for a genuine stationary fractal time series with a unique scaling coefficient ($\alpha = 0.99$, $1/f$ noise). As expected, detrending has little effect on the scaling coefficient of a genuinely self-similar process. Detrending of a composite fractal time series is shown in Fig. 7 (*right*). In the original time series ($\alpha = 0.99$, Fig. 7, *middle*) the central segment of length $N = 39322$ data points (= 30%) was replaced by a subset with $\alpha = 0.79$. Thus, the fractal time series was non-stationary by the scaling coefficient of the composite time series switching between $\alpha \sim 1.0$ and $\alpha \sim 0.8$. The scaling coefficient of the composite time series was 0.97 and thus was only about 2% smaller than that of the original stationary unique fractal time series. Of interest to note is that the scaling coefficient was practically unchanged upon detrending but the SD of the estimates increased for shorter segments. The same behavior is observed for the experimental time series (Fig. 7, *left*) suggesting that the experimental time series had the characteristics of a composite rather than unique fractal process.

By definition, a long-range correlated time series, irrespective of being generated by a composite or unique fractal process, consists of the "drift-like" or "trend-like" fluctuations at all time scales that cannot be removed systematically. In contrast, if the long-range correlations were to disappear upon detrending (expected to be associated with a break-down of the scaling properties) then, most likely, it was only the presence of the trends that led to the "spurious" finding of the long-range correlations. The finding that the scaling coefficient of the experimental series remained unchanged even with higher order detrending is strong evidence that the long-range correlations were not an artifact of non-stationarity. Thus, the observed "drift" (Fig. 2) is an indication of a fractal process and is the consequence of the "built-in" long-range power-law correlations. By the same reasoning, superposition of non-stationary external trends onto control time series with scale-invariant self-similar properties has been demonstrated to have little effect on the scaling properties of the series.[7,12]

4.4 Composite Fractal Time Series and Scaling Properties

In the calculation of the scaling coefficient α by the DFA analysis it is assumed implicitly that the process underlying the fluctuations of the time series constitutes a single fractal process with unique scaling properties. The apparent discrepancy of the results of the experimental 4 hr day-time/night-time time series (Tab. 1) suggests that the fractal properties of the time series were not constant in time but were changing and thus were locally different in the full-length 24 hr series. In order to explore this possibility, composite fractal time series were generated by replacing segments of varied length N of a given time series with known α by segments of another time series with differing known α.

The results are shown in Fig. 8. In the stem series ($N = 2^{17}$, $\alpha = 1.0$) randomly selected segments of length N varied from 10% to 90% of the full length of the series were replaced with segments randomly selected from another series with α varying from 0.9 to 0.5. Ten realizations were calculated for each combination of fractional segmental length and α (for 100 realizations the results were practically the same). It is seen that the scaling coefficient α calculated by the DFA analysis is pushed away from its original value when the fraction of the segmental component is increased, finally approaching that of the series which is used for replacement (Fig. 8, *upper*). An approximately

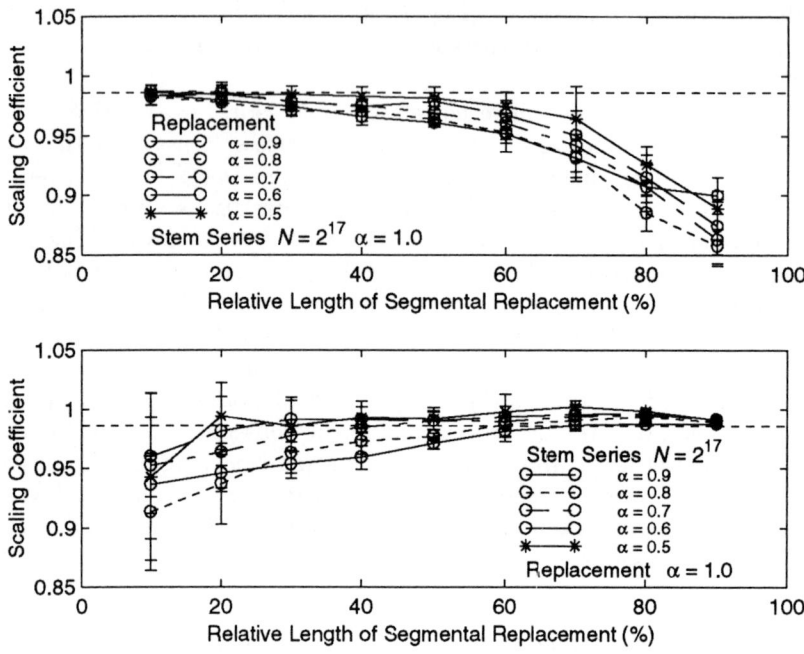

Figure 8: Estimation of scaling coefficient α for composite fractal time series.

mirror image is obtained when fractions of the stem series ($N = 2^{17}$) and varied α (0.9 to 0.5) are replaced by equivalent segments with $\alpha = 1.0$ (Fig. 8, *lower*). Interestingly, the effects are relatively small and even 90% replacement causes α to decrease only by about 15%. Hence, for a composite fractal time series the component with higher α strongly dominates the calculated scaling exponent and the effects are slightly smaller when the difference between the α values is large. Assuming that the experimental subjects had been sleeping for 7.2 hrs (30% of a 24 hr epoch) and their heart rate scaling coefficient during the night-time subepochs was 0.8 and otherwise was 1.0 during the remaining period, the effect is of the order of 2% only. These findings would resolve the conflict that the lower scaling coefficients calculated for the 4 hr night-time subepochs were not reflected in the calculations that were based on the full-length 24 hr time series. On the other hand, these simulations add to the suggestive evidence that the experimental time series had a composite rather than unique structure, i.e. were multifractals.

4.5 Fractal Dimension of Time Series

Formally, the lower 4 hr night-time scaling coefficients as compared to their day-time counterparts are indicative of an *increase* of the fractal dimension of the process underlying the dynamics, or more generally, of an increase of the fractal complexity. The physiological significance of an increased fractal complexity (reduction of error tolerance but increased plasticity, *see Sec. 4.6*) for the control of heart rate during night-time episodes remains unclear. The fractal dimension of the 4 hr day/night subepochs was determined independently by calculations of the *Point Correlation Dimension* (PD2i) which is a modification of the classic Grassberger-Procaccia algorithm for the calculation of the fractal dimension of a deterministic chaotic system and applicable to a stream of non-stationary data.[17,18] The PD2i reconstructs the degrees of freedom (number of independent variables) of the process that generates the time series under study, irrespective of whether the process is deterministic or stochastic and is stationary in time. The PD2i was 2.59 ± 0.30 for the 4 hr day-time subepochs and 3.82 ± 0.43 for the 4 hr night-time subepochs, respectively. Thus, the finding of an increased fractal dimension during the night-time subepochs by the DFA analysis is confirmed by an independent technique of non-linear analysis.

4.6 Long-Range Power-Law Correlated Heartbeat Fluctuations

This study has established that the fluctuations of heart rate constitute an intrinsic feature of a dynamical homeodynamic system emerging from non-linear interactions between different components of the system. The mechanisms underlying the control of heart rate and its fluctuations are acting over a wide range of time scales that are not independent. The fluctuations of heart rate are statistically correlated with fluctuations in the heartbeat intervals thousands of beats earlier ("memory") or the present fluctuations have some impact ahead in time ("futurity") and this correlation decays in a scale-invariant fashion. The behavior of the process is strongly influenced by its entire past history. Furthermore, its memory is dynamic, i.e. the influence of recent events is added to and gradually supersedes the influence of distant events. The influence of the distant past fades very slowly as a power of elapsed time which is characteristic for non-equilibrium systems at the critical point where the much faster exponential decay of the correlations turns into a much slower power-law decay.[20] For $1/f$ dynamics with $\alpha = 1.0$ the decay is logarithmic and even slower than any power of time. For $0.5 < \alpha \leq 1.0$ the influence of the distant past compared with the most recent past increases with increasing α and for $\alpha = 1.0$ present events are approximately equally correlated with events from the recent and the very distant past. Recent events may dominate over short time scales but have a progressively diminishing impact on the longer time scales. Persisting new trends will cause the process to adjust logarithmically with time as progressively more time scales reflect the new trends and supersede the old ones. The fractal process appears to extract trends and condenses information into summaries that, in turn, determine the values of the state variables of the system and influence the somewhat predictable present behavior leaving room for new trends evolving. This scaling behavior calls for an organizational principle for coordinating these time scales which, in physiological conditions, is not badly dominated by any single time scale.

The $1/f$ type of heartbeat interval fluctuations with $\alpha \sim 1$ are in agreement with recent findings in normal subjects at sea level [15] and our results extend these observations by demonstrating stability of $1/f$ fluctuations during exposure to chronic high altitude hypoxia, i.e. altering the pattern of extrinsic environmental factors leaves the intrinsic $1/f$ properties unchanged.

4.7 Physiological Significance of Fractal Dynamics of Heart Rate

West [22] has suggested that a fractal process is more error tolerant and more adaptive to both internal changes and changes in the environment. Here we emphasize that a process with $1/f$ fluctuations ($\alpha = 1.0$) operates at the boundary between stationarity ($\alpha < 1.0$) and non-stationarity ($\alpha > 1.0$). Hence, $1/f$ properties can be interpreted as a "compromise" between two limiting extremes: white noise ($\alpha = 0.5$, completely uncorrelated and unpredictable) and Brown noise ($\alpha = 1.5$, random walk-like Brownian noise). These extremes may be associated with the system's information transfer characteristics and error tolerance. Processes exhibiting the characteristics of uncorrelated white noise ($\alpha = 0.5$) may be associated with high information transfer, high plasticity but low error tolerance while the opposite holds for Brownian noise ($\alpha = 1.5$). In that regard the autonomic nervous system enforces a "compromise" for the heart, i.e. "kicks" away heart rate from these extremes.

In pursuit of this concept the fractal electrodynamics of heart may be summarized in terms of generalized physiological implications and the role of neuroautonomic control of heart rate redefined. There exists an organizing principle for highly complex, non-linear processes that generate fluctuations on a wide range of time scales. The scaling patterns are a consequence of the ordering of the differences, rather than the statistics, i.e. the correlations are produced by the underlying dynamics. The invariant scaling properties (long-range correlations) are consistent with a non-linear feedback system (the autonomic nervous system) that "kicks" the heart rate away from extremes.

This tendency operates on a wide range of time scales, not on a beat-by-beat basis. Lack of a characteristic scale helps prevent excessive mode-locking that would restrict the functional responsiveness (plasticity) of the control system. In a classical (non-fractal) process operating on a single dominant time scale any additional scale introduced by fluctuations has a devastating effect on the process. Scale-invariant correlations ($1/f$ noise) offers the best compromise between efficient information transfer and immunity to errors on all time scales.

A fractal process underlying heart rate dynamics is essentially unresponsive to error and tolerant of the variability in the physiological environment (e.g. hypoxia). This error tolerance results from the broad-band nature of scale size, any scale introduced by error is already present in the system. A fractal process is preadapted to errors or variations in the environment. Hence, the heart is preadapted to systemic (arterial) hypoxia as demonstrated by the lack of any adaptation during chronic exposure. The long-range correlations of the heartbeat series would seem to indicate that the mechanisms of neuroautonomic control drive the system away from a single steady state. The classical theory of homeostasis according to which stable physiological processes readjust to maintain "constancy" should therefore be revisited to account for the dynamical non-equilibrium behavior.

Acknowledgments

Supported by Swiss National Science Foundation, Grant n° 32-040397.94 and by American National Institutes of Health, Grant n° NS27745.

References

1. P. Bak and M. Creutz, in *Fractals in Science*, eds. A. Bunde, S. Havlin (Springer, New York, 1994).
2. J. Beran, *Statistics for Long-Memory Processes* (Chapman & Hall, New York, 1994).
3. S.V. Buldyrev *et al.*, *Biophys. J.* **65**, 2673 (1993).
4. L. Glass and C.P. Malta, *J. Theor. Biol.* **145**, 217 (1990).
5. B. Grassi *et al.*, *J. Appl. Physiol.* **82**, 1952 (1997).
6. J.M. Hausdorff *et al.*, *J. Appl. Physiol.* **78**, 349 (1995).
7. J.M. Hausdorff *et al.*, *J. Appl. Physiol.* **80**, 1448 (1996).
8. M. Meyer *et al.*, *Applied Cardiopulmonary Pathophysiology* **4**, 213 (1992).
9. M. Meyer *et al.*, *Integr. Physiol. Behav. Sci.* **31**, 289 (1996).
10. D.A. Murphy *et al.*, *Am. J. Physiol.* **266**, R1127 (1994).
11. M.S. Keshner, *Proc. IEEE* **70**, 212 (1982).

12. C.-K. Peng *et al.*, *Phys. Rev.* E **47**, 3730 (1993).
13. C.-K. Peng *et al.*, *Physical Review Letters* **70**, 1343 (1993).
14. C.-K. Peng *et al.*, *Phys. Rev.* E **49**, 1685 (1994).
15. C.-K. Peng *et al.*, *Chaos* **5**, 82 (1995).
16. G. Samorodnitsky and M.S. Taqqu, *Stable Non-Gaussian Random Processes: Stochastic Models With Infinite Variance* (Chapman & Hall, New York, 1994).
17. J.E. Skinner *et al.*, *Am. Heart J.* **125**, 731 (1993).
18. J.E. Skinner *et al.*, *Integr. Physiol. Behav. Sci.* **29**, 217 (1994).
19. J.E. Skinner and J.Y. Kresh, in *Nonlinear Techniques in Physiological Time Series Analysis*, eds. G. Mayer-Kress, H. Kants, J. Kurths (Springer, New York, 1996).
20. H.E. Stanley, *Introduction to Phase Transitions and Critical Phenomena* (Oxford Univ. Press, New York 1971).
21. R.F. Voss, in *The Science of Fractal Images*, eds. H.O. Peitgen, D. Saupe (Springer, New York, 1988).
22. B.J. West, *Ann. Biomed. Eng.* **18**, 135 (1990).

EPISTEMOLOGICAL AND TREATMENT IMPLICATIONS OF NONLINEAR DYNAMICS

A. H. STEIN

200 West 70th Street, Suite 9K, New York, NY 10023, USA
phone: 01+212.362.2559; fax: 01+914.424.4416; email: astein@necet.edu

Abstract

The treatment implications of understanding mind as solely epiphenomenal to nonlinearly founded neurobiology are discussed. G. Klimovsky's epistemological understanding of psychoanalysis as a science is rejected and treatment approaches integrating W. R. Bion's and D. W. Winnicott's work are supported.

1 G. Klimovsky's Psychoanalytic Science

In "Epistemological Aspects of Psychoanalytic Interpretation,"[1] G. Klimovsky argues for Psychoanalysis' scientific status as follows: since, by their nature, sciences operate according to the rules of correspondence, "if A then B," and since Psychoanalysis posits rules that associate manifest with latent content, Psychoanalysis is a science, Q.E.D.[2] He advances his argument by comparing the training and consequent behavior of biologists and psychoanalysts. Like the microscopist, who, having learned correspondence rules that associate perceived images within a microscope's objective with actual objects on his slide, can subsequently "read" the perceived image as though it were the object, itself, the psychoanalyst, having learned correspondence rules that associate manifest with latent content can just as confidently "read" latent content when presented with manifest content.[3]

As Klimovsky apparently intends them, then, both cases can be portrayed by the differentiable function, $X = a(\lambda) + b$, where $\frac{dX}{d\lambda} \in \Re^n$, illustrated in Figure 1. In his microscopy example, if x represents the value of any actual object's arbitrary dimension, X, and λ, the value of that dimension's image as seen in the microscope's objective, a is a scaling function. For Psychoanalysis, if x represents the value of any continuous psychoanalytic factor, X, such as ego, and λ, particular measurable behaviors ostensibly associated with some level of development—the psychoanalytic equivalent of time—the scaling function, a, becomes a growth function that associates the two.

In the latter case, the line, $X = a(\lambda) + b$, has generally come to portray a universal developmental trajectory that associates monotonically increasing psychic functions with particular ages. The existence of such a universal trajectory allows analysts to identify pathogenic loci at which an individual

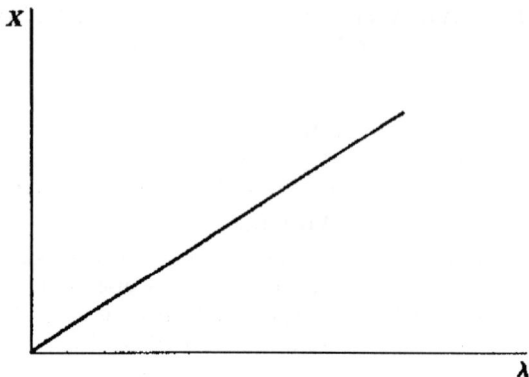

Figure 1: The linear function $X = a(\lambda) + b$.

trajectory may have diverged from the universally normative one. This, in turn, allows them to formulate and deliver interpretive corrections, that can lessen that divergence.

2 Closed, Time-differentiable, Nondissipative Systems

Differentiable functions make assumptions about reality that, while generally irrelevant to microscopy, are fundamental to psychoanalytic metapsychology and treatment.[4] In time-differentiable reality, for example, elements constituting systems are understood to be independent of each other in that they do not exchange energy or information with their environments; under these circumstances, there is one and only one value of X for each value of λ. Given those assumptions, differentiable equations are generally used to describe dynamically closed, continuously time-differentiable, nondissipative systems, for which those assumptions hold.

Within such systems, defining λ as time allows that the same value of λ gives rise to the same value of X whether it is approached from the left, i.e., from "earlier" or from the right, i.e., from "later;" time is reversible. It is precisely these ideas that constitute the necessary conditions for psychoanalytic regression and repair through interpretation: pathogenic loci can be approached and repaired *via* interpretation from the direction "earlier←later" even though they were originally laid down in the direction "earlier→later."

3 Open, Non-time-differentiable, Dissipative Systems

Living systems, however, rarely operate according to such time-differentiable dynamics. They are necessarily open, nonlinear, and dissipative. They exchange energy and information with their environments; their constituents are interactive and, therefore, they have multiple potential values of X $(x, x', x'', ...)$ for any given λ. X's actual value appears only in the moment of observing it, as a joint function of the system's initial conditions and particular history. This situation is illustrated in Figure 2.

There, the fixed points of a nonlinear dynamical system are displayed as a function of the parameter, λ. To understand its implications, assume that

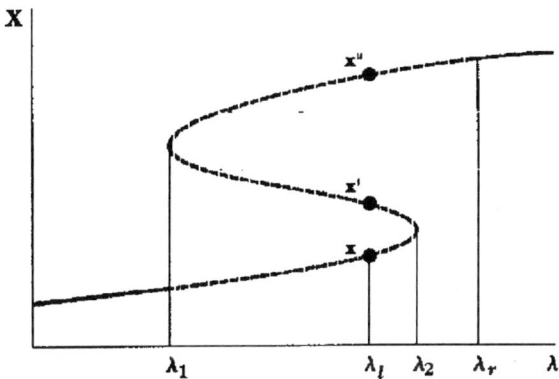

Figure 2: The fixed points of a nonlinear dynamical system as a function of the parameter, λ.

λ is varied slowly so that the trajectory of the system remains close to the associated fixed point. Assume also that the system is on the lower most portion of the curve and that λ is increasing. The value, λ_l, is allowed to increase to λ_2 and then slowly decreased back to λ_1.

As λ_l decreases, the system can "choose" among an infinite number of fixed points, three possible ones being, x, x' and x'', all of which can differ markedly from its original course. If λ is interpreted as time, then the trajectory from λ_1 to λ_2 is not necessarily the same as from λ_2 to λ_1. While the characteristics of differentiable functions hold for $\lambda < \lambda_1$ and $\lambda > \lambda_2$, between those values time is not necessarily reversible, history is not preserved and regression will neither provide us with "lived" nor objective, historical experience. This is so because when they are approached backwards, unpredictable areas, within which many values of X can exist for a given λ, emerge alongside areas wherein single values of X exist for each λ. The latter case can, in fact, be understood as a special case of the former.

3.1 Implications of Open, Non-time-differentiable, Dissipative Dynamics for Psychoanalysis

While psychogenic loci can sometimes be approached for repair *via* regression, then, sometimes they cannot. Unless one is prepared to accept a dualistic position which asserts that psychic reality operates according to laws different from material reality, this has significant implications for Psychoanalysis, as received. Among other things, since one can never know when he is within which area, one must doubt the certainty and meaning of analytic interpretation, to the extent that interpretation is based upon developmental or genetic correspondence rules. W. R. Bion and D. W. Winnicott were aware of these issues.

W. R. Bion

To account for the unpredictability of psychic contents, in his model for thinking, Bion[5a], proposed that thought arises as a state transition, Ps→D, that occurs within the psychic stuff, β: Ps→D describes one half of Ps↔D, the psychic oscillation, fragmentation↔organization. D's actual form, D_i (the particular x value in the example, above), is jointly determined by two factors, α-function and the "selected fact," s. α-function, is an evolving, time-differentiable "function of the psyche" (represented in Figure 1 and in Figure 2, to the left of λ_1 and to the right of λ_2) that, by virtue of its accumulating history, possesses all the properties of the differentiable function, $X = a(\lambda) + b$; s,[6] on the other hand, is a non time-differentiable, metaphysical element given up by 'O', "...ultimate reality...absolute truth, the godhead, the infinite, the [unspeakable] thing-in-itself"[7] which Eigen[8] envisions as arising unpredictably

---

[a]Trained as a physicist, Wilfred Bion attempted to reckon psychoanalytic metapsychology formally. He sought, for example, to map psychoanalytic ideas onto biological ones as a way to repair the consequences of Freud's having severed psychodynamics (dynamics of the "sensual") from material dynamics (dynamics of the "extra-sensual").

This aspect of his work often led him to speak in quasi-mathematical and physically dynamical language about metaphysical and psychoanalytic entities including Kantian β-elements, psychoanalytically defined "depressive" and "paranoid-schizoid" positions (D and Ps), undefined α-functions and "selected facts," given up by transcendent "O," the Kantian *ding-an-sich*, through Faith. In spite of these superficial confusions, Bion generally separated the physical from the metaphysical aspects of his own theory making it possible for us, today, to analyze the latter, at least functionally, to determine if and how they can be explained in terms of the former, given our new mathematical tools generally applied to nonlinear dynamic systems.

and discontinuously—as an act of Grace through Faith.[9] It is ahistorical and possesses none of the properties of differentiable functions. Its appearance fractures α's continuity at unpredictable times and in unpredictable ways while, at the same time, it provides a metaphysical "armature" around which α-function can "pick up the fractured pieces" and organize a particular, novel thought, D_i, i.e., provide the actual value, $x \equiv D_i = \alpha_j(Ps, s)_{Faith\ in\ O}$.

Between consecutive D_is (a consequence of the reverse state transition, $Ps{\leftarrow}D_i$) lies the fractured state Ps, the "pieces," a boundless state of unknowing, wherein thoughts do not exist. Subsequent reappearances of D_i, then, phenomenologically rescue one from non-knowing by making the world comprehensible again, at least for the moment, until it vanishes in the next state transition into Ps.

Bion held out the possibility that, by "linking" with a patient and "becoming 'O'," ostensibly becoming a "channel," of sorts, for 'O,' the analyst can intuit the patient's psychic process with sufficient clarity that, *in loco* 'O,' he can provide a properly selected fact, in the form of an interpretation. It, then, can provide the core for a novel D_i and, thereby, organize the patient's fragmented experience. He was also deeply aware, however, both of s's inherent unpredictability and of the inevitable anxiety that arises from linking–with–another–without–thoughts, i.e., without intentions. This led him to caution analysts to approach each analytic encounter "without memory, desire, or understanding,"[10] i.e., to faithfully await s's appearance rather than to attempt to force its presence by imagining what it might be if it ever did arrive.[b]

D. W. Winnicott

D. W. Winnicott addressed unpredicatbility in the analytic situation and its consequences for treatment differently. Rather than assuming that the analyst could function *in loco* 'O' if only he could "become 'O'" *via* linking, in the first

---

[b]A problem arises when both analyst and patient experience the phenomenological state, Ps, at the same time; both may become anxious at the existential intimacy of moments-without-knowledge and one or both may defensively attempt to abdicate Faith and to force the production of a thought by inventing an s. Bion[11] described this possibility as follows: "By agreeing with the [analyst's] interpretation [the patient may] hope that the analyst will be inveigled into a collusive relationship [with him] to preserve ["knowledge" regarding their relationship thus]...preventing the inception to $T\alpha \to T\beta = K{\to}O$ [transformation in "O"]." Freud[12] had described the situation in terms of a patient's "meaningless" acknowledgment of "the correctness of the [analyst's] construction...[which]...may be convenient...in order to prolong the concealment of a truth that has not been discovered." While both of these caveats refer to the patient's anxiety, the analyst can collude with the patient's forced understanding out of identical anxiety.

place, Winnicott removed that possibility from his theory. Instead, he posited that analytic interaction occurs within a coadapting kaleidoscopic realm, that he called "transitional space." Therein, interaction takes a particular form that he called "play."

He[13] illustrated its essence by describing his interaction with a young patient, Iiro. I repeat its opening without Winnicott's commentary, which underlines its inherent difficulties; I will address them, presently.

> I said: "I shut my eyes and go like this on the paper and you turn it into something, and then it is your turn and you do the same thing and I turn it into something." I made a squiggle which turned out to be of the closed variety. He quickly said: "It's a duck's foot." ...I did a drawing with the webbed foot of a duck delineated...He now chose to draw and he produced his own version of the webbed foot of a duck....Next I did an open squiggle which he immediately turned into a duck swimming in the lake...He now did [a] squiggle and he turned it into a horn...I made a squiggle and he quickly turned it into a dog...He made a squiggle which I turned into a question mark. This was evidently not what he had in mind as he said: "It could have been a hair." ...I made his squiggle into a rather awkward-looking swan...."[14]

In transitional space, both Winnicott and Iiro selected each other's behavior by noninterpretive interactions using squiggles. Ostensibly, both interacted as equals in the existential moment. Abdicating the analyst's traditional authority, Winnicott seemed to approach this interaction without memory desire or understanding, as Bion had encouraged.

Yet, in his commentary, wherein he reported his internal experience of his interaction with Iiro, Winnicott pointed to an inherent problem applying his theory to practice. When Iiro responded to Winnicott's "closed" squiggle with the statement that "It's a duck's foot," rather than responding to it as he had responded to squiggles, he followed Klimovsky: he *read* Iiro's "manifest" comment about ducks as a "latent" "...wish...to communicate with me on the subject of his disability."[15] As a result, he *de facto* halted play in the existential moment and, instead, "...lean[ed] back and wait[ed]..." until Iiro "was...able to acknowledge and cope with the idea of his syndactyly."[16]

Winnicott later understood this behavior as a consequence of two factors: one was related to his having to "tolerate the existence of two contrary trends in oneself,"[17] (to retain a modicum of analytic objectivity in his practice while remaining truly "playful," in Bion's sense of participating in a "transformation in 'O'"[18]), and to applying his theory, in general.[19]

4 The Epistemological Problem of "Knowing" in Nonlinearly Founded Psychoanalysis

To return to the general epistemological problem, experiences of "reading" and "knowing" are inherent to play, itself. When thoughts seemingly appear out of nowhere *via* the state transition Ps→D$_i$, their nature is to be true. Even if their content is "I'm not sure this is true," or "this must be false," those declarative statements are the truth-of-the-moment. They are, in that way, like squiggles.

Our problem, then, concerns what to do with our truth-of-the-moment "mental squiggles" when they seem to us to apply to another's experience, especially when we are analysts with all authority inherent in that position *vis a vis* our patient. Bion would have us guard against narcissistically applying our truth, as though it were theirs, to them, i.e., using analytic authority as a bat. But, while nobly stated, this is not always an accomplishable task: fear of intimacy and narcissism are ubiquitous. Even if they were not so, the nature of nonlinear systems prevents us from ever knowing when we are operating in a differentiable or non-differentiable slice of reality at any given time, no matter what the contents of our thoughts may indicate.

Winnicott's solution, to understand our interactions as play within transitional space, is rife with difficulties, as well. In Winnicott's as well as Bion's case, as analysts, we can "aintentionally" use the instantaneous contents of our thoughts to protect ourselves, by either halting play and destroying transitional space without our patient's awareness or by forcing a fact of our own selection upon both of us to protect ourselves from the anxiety inherent in not knowing; in Winnicott's language, both patient and analyst can collude in denying that fact by jointly acting "as if" the transitional space remained intact. Here, following Freud and Bion, above, Winnicott spoke of the general problem in terms of "interpretations...that the patient accept[s] because of awe [I would include fear, that] turned out in the end to be collusive defenses."[20]

The problems that arise when applying either Bion's or Winnicott's system to practice do not appear to be inherent failures within the respective systems. Instead, they seem to emerge from our own fears of intimacy and of ignorance when applying them. To the extent, that we, as analysts, can manage those fears, however, an integration of Bion's and Winnicott's systems seems most compatible with a Psychoanalysis founded upon nonlinear dynamics.

5 Conclusion

Given the uncertainties intrinsic to nonlinear dynamics and the complexities of interpersonal relating, there exists no neat answer to how analysts should "best" behave with patients. Indeed, even if we accept this as a meaningful question, nonlinear dynamics precludes our formulating prescriptive answers that function other than to protect us from living in the flux of the existential moment, i.e., suffering "transformation in 'O'." Further, human fears of intimacy and the epistemological uncertainties inherent in interpersonal relating constrain Bion's injunctions and Winnicott's reframing, from viably solving Klimovsky's "reading" problem. Indeed, from the perspective of nonlinear dynamics, analyst's "readings" of patient's behavior are better understood as political responses[21] to patients-unpredictably-behaving-in-the-moment than as the objective scientific readings that Klimovsky envisions.

6 References

1. G. Klimovsky in *Fundamentals of Psychoanalytic Technique*, ed. R. H. Etchegoyen, 471-493 (Karnac Books, London, 1991).
2. *Ibid.*, 478.
3. *Ibid.*, 479.
4. A. H. Stein in *Nonlinear Dynamics in Human Behavior*, eds. W. Sulis and A. Combs, 256-275 (World Scientific, Singapore, 1996).
5. W. R. Bion, *Attention and Interpretation* (Tavistock, London, 1970).
6. *Ibid.*, 42.
7. W. R. Bion, *Learning From Experience*, 26 (Heinemann, London, 1962).
8. M. Eigen, *Int. J. Psych.-Anal.*, **62**, 413-433 (1981).
9. W. R. Bion, *Attention and Interpretation*, 32-3 (Tavistock, London, 1970).
10. *Ibid.*, 59.
11. W. R. Bion, *Transformations*, 160 (Heinemann, London, 1965).
12. S. Freud in *The Standard Edition of the Complete Psychological Works of Sigmund Freud*, ed. and trans. J. Strachey, **23**, 262 (Hogarth, London, 1937).
13. D. W. Winnicott, *Therapeutic Consultations in Child Psychiatry* (Basic Books, New York, 1971).
14. *Ibid.*, 12-16.
15. *Ibid.*, 13.
16. *Ibid.*, 16.
17. *Ibid.*, 9.

18. W. R. Bion, *Transformations*, 155-157 (Heinemann, London, 1965).
19. A. H. Stein in *Nonlinear Dynamics, Psychology and Life Sciences* (forthcoming).
20. D. W. Winnicott, *Ibid.*, 10.
21. A. H. Stein, *Splitting, Induction and the Politics of Borderline Object Relations* (unpublished, 1989).

THE SIX FUNDAMENTAL CHARACTERISTICS OF CHAOS

AND THEIR CLINICAL RELEVANCE TO PSYCHIATRY:

A NEW HYPOTHESIS FOR THE ORIGIN OF PSYCHOSIS

GARY BRUNO SCHMID, PH.D.
Psychologist / Psychotherapist SPV/ASP
General Psychiatry
Cantonal Psychiatric Clinic Rheinau
CH-8462 Rheinau, Switzerland

140

ABSTRACT

Underlying idea: A new hypothesis about how the *mental state of psychosis* may arise in the brain as a "linear" information processing pathology is briefly introduced. This hypothesis is proposed in the context of a complementary approach to psychiatry founded in the logical paradigm of chaos theory. To best understand the relation between chaos theory and psychiatry, the semantic structure of chaos theory is analyzed with the help of six general, and six specific, fundamental characteristics which can be directly inferred from empirical observations on chaotic systems. This enables a mathematically and physically stringent perspective on psychological phenomena which until now could only be grasped intuitively: Chaotic systems are in a general sense dynamic, intrinsically coherent, deterministic, recursive, reactive and structured; in a specific sense, self-organizing, unpredictable, nonreproducible, triadic, unstable and self-similar. To a great extent, certain concepts of chaos theory can be associated with corresponding concepts in psychiatry, psychology and psychotherapy, thus enabling an understanding of the human psyche in general as a (fractal) chaotic system and an explanation of certain mental developments, such as the *course* of schizophrenia, the *course* of psychosis and psychotherapy as chaotic processes.

General overview: A short comparison and contrast of classical and chaotic physical theory leads to four postulates and one hypothesis motivating a new, dynamic, nonlinear approach to classical, causal psychiatry: Process-Oriented PSYchiatry or "POPSY", for short. Four aspects of the relationship between chaos theory and POPSY are discussed: (1) The first of these, namely, Identification of Chaos / Picture of Illness involves a definition of Chaos / Psychosis and a discussion of the 6 logical characteristics of each. This leads to the concept of dynamical disease (definition, characteristics and examples) and to the idea of "psychological disturbance as dynamical illness". On the one hand, it is argued that the *developmental course of psychosis is chaotic*. On the other hand, we propose the *hypothesis that the mental state of psychosis may be a linear information processing pathology*. (2) The second aspect under discussion is the Assessment of Chaos / Diagnosis of Illness. In order to better understand how POPSY research treats this aspect, we take a look at the 3 different classes of (non-quantum) motion as models of 3 different possible courses of illness and outline present-day methods available for the quantitative assessment of chaotic (fractal) motion. (3) The third aspect, namely, Prediction of Chaos / Prognosis of Illness considers how each of these 3 classes of motion implies a different way of looking into the future: linear-causal, statistical and nonlinear-fractal, respectively. (4) The fourth aspect of the relationship between chaos theory and POPSY, Control of Chaos / Treatment of Illness, is shown to have certain implications to complementary medicine. This paper completes with a short summary, conclusion and a closing remark.

KEY WORDS: chaos, dynamical disease, psychiatry, psychology, psychosis, psychotherapy, schizophrenia

1 INTRODUCTION

<u>Chaos vs. Classical Theory:</u> Chaos Theory - like its theoretical counterpoint: classical mechanics - offers certain metaphors and pictures which deepen our mathematical-physical understanding of everyday life. For example:

STRUCTURES: Classical mechanics enables us to describe the structure of a palace; chaos theory, the structure of a fern leaf.

MOTION: Classical mechanics explains the trajectory of a rocket; chaos theory, the meandering of a butterfly.

SYSTEM BEHAVIOR: Classical mechanics helps us understand the dynamics of our solar system; chaos theory, the dynamics of a juggling act.

These examples speak for themselves: Our macroscopic observations of the inorganic world can be fairly well explained by classical mechanics, i.e., with a linear, reductionistic "building block" world picture; those of the living world, with chaos theory, i.e., with a nonlinear, nonreductionistic "internetted" world picture.

<u>4 Postulates and 1 hypothesis.</u> Consider now the following ideas.
The juggling of 3 objects depends not only upon the force of gravity and the gravity-defying muscle power of a juggler but, rather and decisively, also upon the space-time situation of the dynamic interplay of these forces.

The flourishing of a plant depends not only upon the presence of light, air, water and earth but, rather and decisively, also upon the space-time situation of the rhythmic interplay of these elements.

In this spirit, I would like to motivate my discussion with 4 postulates and 1 hypothesis.

POSTULATE I: The pathological effect of a damaging influence depends not only upon the presence of a biological predilection, a psychological sensitivity or an unfortunate sociological constellation but, rather and·decisively, also upon the space-time situation of the dynamic interplay between the respective vulnerable elements.

QST: Which "sociopsychobiological clock" must we pay attention to in order not to get ill?

POSTULATE II: The healing effect of a therapeutic treatment depends not only upon medical aid, psychological capability of the therapist or social help but, rather and decisively, also upon the space-time situation of the dynamic interplay between the respective supportive elements.

QST: Which "sociopsychobiological clock" must we pay attention to in order to remain healthy or to get well again?

POSTULAT III: The complex dynamic interplay between detrimental influences, vulnerable elements, therapeutic treatment and supportive elements is decisive for human health.[a]

QST: Which methods should we employ in order to set the above-mentioned "clocks" and to direct their complex dynamic interplay for optimal human well-being?

POSTULATE IV: Chaos theory offers mathematical methods according to which the above-mentioned "clocks" can be systematically searched for, recognized, understood and set.

HYPOTHESIS: The interplay of chaos theory and psychiatry leads us through the threshold of a new approach to psychiatry.

Insofar as we may use chaos theory to help us cross this threshold, we are professing a kind of biological psychiatry in the truest sense of the prefix "bio", namely, with an emphasis upon living processes (as opposed to mechanical or biochemical conditions). I would therefore like to call this new, dynamical-nonlinear approach: Process-Oriented PSYchiatry or "POPSY", for short.

In order to differentiate between the terms psychology (psychological etc.), psychiatry, (psychiatric etc.) and psychotherapy (psychotherapeutic etc.), I would like to suggest, for the purposes of this paper, the following simplified definitions: Psychiatry looks at the patient from a classifying standpoint; Psychology is observation of the patient from an interpreting perspective. Psychotherapy, in contrast, is view of the patient through empathic identification with him or her (cf. [3]). The importance and implications of this distinction have been followed through in detail elsewhere in a related work [21].

2 CHAOS THEORY AND POPSY

A classifying psychiatry is based upon causal thinking: the short- and long-term future is contained entirely in the *temporal course of preceeding sequential events in physical space*. A process-oriented psychiatry is based upon synchronistic thinking: the immediate future is contained in a *cluster of nonsequential events from the recent and distant past within a dynamical state space*. Although seemingly

---

[a]An example from football: Even if the internal coordination of the goalkeeper functions perfectly well, he still won't be able to sucessfully defend the goal if he "gets off on the wrong foot".

nonassociated, these points possess long-range correlations as evidenced by the cluster's *Gestalt*, a so-called "attractor", in this state space (see below).

Synchronistic thinking offers an alternative to a reductionist world picture which relies primarily upon traditional methods of analysis such as statistics, multivariate methods, spectral analysis etc. With the help of suitable nonlinear dimensional analyses, a synchronistic world picture focuses upon the general structure and complexity of the behavior of the "generator" of the observed (anthropological) system.

A process-oriented psychiatry introduces a new area of research offering: (a) the objective *time-series evaluation* of a disturbance and its treatment (single-case, course of illness research); (b) insight into the complexity of a disturbance, i.e., into the *number of variables / degrees of freedom*, which fundamentally influence the time-Gestalt of the disturbance and which, together with certain empirically induced conditions, may allow for *mathematical models* of the disturbance (outbreak, course and remission or chronification); (c) Potential for *short-term, individual prognosis;* (d) Potential for *therapy* in the sense of "chaos control"

Points (c) and (d) deserve special attention. It is reasonable to assume that several early attempts or "virtual disturbances" generally preceed the full outbreak of any psychiatric illness resulting from a complex net of nonlinear sociopsychobiological influences. In the case that such virtual disturbances leave a reliable, unique signature in the reconstructed state space describing the dynamics of the system, a time series protocol of healthy episodes may evidence sporadic "distortions" (early warning signals) which, on an individual basis, reliably warn of an impending pathological "jump". Accordingly, a preventative therapeutic intervention taken early enough during this development could hinder such a "jump" and all the ensuing pathological consequences. These considerations form the basis for the application and utility of "nonlinear forecasting" in the domains of psychiatry, psychology and psychotherapy: Nonlinear methods are already being used to warn individuals about the probability of an impending heart attack on the basis of weak, isolated ECG (electrocardiogramm) outpost symptoms in the absence of any obvious manifest signs of functional disturbance to the heart [16]. Similarly, the methods of nonlinear forecasting, in particular with the help of $f(\alpha)$-spectra and associated picture processing techniques, may make possible *reliable, individual prognoses* of otherwise apparently random and unstable psychopathological processes.

Let us now risk a general comparison and contrast: classifying versus process-oriented psychiatry. We will look at (a) the associated structure of the assumed "unit of illness" sought for in each case; (b) the respective sources of information assumed to be necessary for each search; (c) the respective pictures of illness underlying each search.

The structure sought for in a classifying psychiatry is the clinical picture of the personality and illness necessary to make a diagnosis. The source of information necessary for this search arises from sufficiently detailed and complete observations from various sectors of the afflicted individual's life, e.g. work, relations to unknown third persons, to colleagues, friends, family etc. (direct and third-person interviews). *In this case, health is understood to lie primarily in regular, periodic functioning (homeostasis).* Deviations from standard values (e.g. body temperature, number of absences from work per month, etc.) indicate illness.

The structure sought for in a process-oriented psychiatry is a "portrait in state space", a cluster of long-range correlated points forming a Gestalt called an "attractor". The source of information required to find this structure can arise from sufficiently differentiated observations of only *one* well-defined symptom or *one* certain sporadically repeating situation in the life of the afflicted individual, e.g. the occurence of a single psychotic symptom (when and how strong?). Such time-series observations enable one to reconstruct the associated "state space" and to deduce from this, via a corresponding time-series analysis, the minimum number of degrees of freedom of the respective generator of the disturbance. *In this case, health is understood to lie primarily in the high variability of functioning.* Restricted variability (e.g. exclusively neurotic or psychotic behavior) - or suddenly occuring regularities (e.g. obsessive compulsions, depressive monotonicity, affective flattening in schizophrenia etc.) - point to illness.

The following table outlines

2.1 Four aspects of the relationship between chaos theory and process-oriented psychiatry

Mathematical / physical Problem:	Psychological / psychiatric Problem:
IDENTIFYING chaos "Analysis in state space"	... the disturbance "Picture of illness / Anamnesis"
QUANTIFYING chaos "Attractor"	... the disturbance "Diagnosis"
PREDICTING chaos "Forecasting"	... the course of disturbance "Prognosis"
CONTROLLING chaos "Attractor stability / Complexity"	... the disturbance "Treatment / Therapy"

Example: Henon Attractor   Example: Schizophrenic Psychosis

Each aspect has in turn 3 different facets:

1.   Formal logical basis supported by theoretical (mathematical-physical) considerations.

2.   Realisation of these considerations withing the framework of concrete mathematical, physical and clinical projects.

3.   Practical consequences with regard to the quantitatively relevant application of these projects in actual clinical practice.

Let us now consider each aspect in turn onhand the example of psychosis, in particular, within the context of schizophenia.

2.2 First Aspect: Identification of Chaos / Picture of Illness

This aspect raises the question: *"What is chaos / psychosis?"* To help answer this question, imagine an actual encounter with a psychotic patient and compare your imagination of this patient's behavior with your remembrance of a juggling act. The recognition of chaotic behavior (e.g. turbulence) is based upon long-term observations of the system onhand the *form* and not the time-span of appearance of certain, obvious phases. The recognition of a psychopathological disturbance (picture of illness) is similar from the standpoint of process-oriented psychiatry. We will see below that the course of a schizophrenic psychosis displays the same six formal characteristics as the behavior of a chaotic system. For this reason, we will be able to understand schizophrenia as a so-called "dynamical disease" and to view the picture of illness from the perspective of its long-term course (=schizophrenic process), namely, from the same perspective which a mathematician uses to understand a mathematical attractor. The problem still remains, however, for psychiatric research to show whether or not the schizophrenic course of illness possesses a generic form from which a reliable individual diagnosis can be made onhand a chaostheoretical analysis of the corresponding time series of one or another schizophrenic symptom.

Let us now propose the following

2.2.1 Working definition of chaos:

CHAOS: *A kind of order between the poles of randomness and periodicity. The behavior of a chaotic system has six formal logical characteristics and accordingly displays "non trivial" (so-called "long range") correlations which only become evident between two or more modes of behavior over an extended period of time without possessing any particular, characteristic time scale.*

(As a consequence of such long range correlations, chaotic systems have an enhanced adaptability to sudden, unexpected, external influences. As a consequence of having no characteristic time scale, chaotic dynamics are generally fractal.)

There is no universally accepted definition of chaos in the scientific literature [11]. For this reason, I have already proposed in another publication [21] the following six basic phenomenological characteristics of chaotic processes which, taken together, are sufficient for the definition of the corresponding system - actually and more exactly: for the *behavior* of the system - to be chaotic. These can easily be dichotomized into two parallel classes which I would like to call (a) "general" and (b) "special". The general attributes also characterize mathematical-physical systems of a non-chaotic nature. The special attributes, however, represent more specific characteristics which are particular to chaotic systems.

General attributes of chaos	Special attributes of chaos
(1a) dynamic	(1b) self-organizing
(2a) intrinsically coherent	(2b) unpredictable
(3a) deterministic	(3b) non-reproducible
(4a) complexly recursive	(4b) triadic
(5a) reactive	(5b) unstable
(6a) structured	(6b) selfsimilar

Each of the six special fundamental characteristics corresponds to one of the six general characteristics in the sense of a further refinement of the general concept. Although each of the second and third pairs of characteristics seem to posses respective elements which oppose one another, namely, "How can a system be both intrinsically coherent and yet at the same time unpredictable?" and "How can a system be both deterministic and yet not reproducible?", these apparent contradictions lie in our customary linear-causal way of thinking and not in the respective pair of characteristics themselves. Indeed, the corresponding concepts belonging to each pair of fundamental characteristics should be understood neither as opposites nor as synonyms but, rather, as rank-ordered (from "general" to "specific") correlates. Finally, the six pairs contain (from my point of view and lacking a formal proof) sufficient heuristic, empirical criterion according to which a system is, in a strict mathematical-physical sense, chaotic.

Strictly speaking, the six characteristics of each class - general or specific - are neither logically nor mathematically independent. They overlap and interrelate in sundry ways, mutually determine each other in part and may, in one manner or another, be more or less dependent. In particular, the following four interplays between characteristics are especially important [21]: (1) The first, second, third, fourth and sixth pairs of characteristics contain several aspects of *determinism*, namely, that the changes of state of a system organize themselves (self-organisation) recursively (recursivity) while manifesting a particular kind of inner regularity (triadicity and self-similarity) and thereby structuring themselves (intrinsic coherence and structuredness) according to certain laws (determinism); (2) The second and third pair of characteristics contain two aspects of *autonomy* in the sense of the impossibility to make long-term predictions of arbitrarily good (a) precision, no matter how many time series measurements may be at one's disposal (unpredictability), and (b) reliability, no matter how exactly the initial conditions may be reproduced (nonreproducibility); (3) The third and fifth pair of characteristics encompass two different aspects of *spontaneity* insofar as dramatic changes in behavior (lability) may arise from reactions to very small changes in initial conditions or in otherwise constant system parameters (nonreproducibility); (4) Finally, the third, fourth and fifth pair of characteristics present three different aspects of *nonlinearity* as has been discussed in detail elsewhere [21] in the discussion to the fourth special characteristic, triadicity. The list of such interplays can be easily extended, for example, in consideration of concepts like "long-range correlation", "causality" and many more.

"Chaotic" on chaos is the interplay of contradictions: A self-organizing determinism in a dynamical state space in contrast to a high degree of disorganisation in space-time behavior. Perhaps we may speak here of a "lawful disorder". As a consequence, deterministic laws may underlie "confused", irregular socio-, psycho or physiological data: A nonreproducible helter-skelter in human behavior or irregularities in repeated measurements does or do not necessarily indicate inadequate, erroneous observations or randomness. There remains a third possible explanation: The behavior is chaotic (in the mathematical-physical sense).

In this way, chaos turns out to be a new class of deterministic, dynamical behavior complementary to the canonical modes of behavior: random, periodic and constant.

If the behavior of a system is chaotic, a certain lawfulness underlies the associated motion and this lawfulness can be discovered. The mathematical description of chaotic lawfulness is in turn an important first step in the direction of *limited, individual prognosis* [6], [27] *and control* [19] of illness.

Individual prognosis (as opposed to statistical forecasting on the basis of a canonical population) and treatment is especially important in the field of medicine. Indeed it is well known that the laws of chaos underlie certain somatic disturbances of the human organism for which neither injury nor poisoning, nor any kind of foreign agent (bacteria, virus, macrophage etc.) or genetic development is responsible. Such functional disturbances have been called "dynamical diseases" in the literature [1], [7], [13], [14]. On the other hand, it has become clear that chaotic behavior can be very important for the healthy functioning of certain organs such as the heart [9], [16] or the brain [25] (see also [18]). Insofar as certain modes of behavior, whether healthy or pathological, could be chaotic in the mathematical-physical sense, we are encouraged to modify many of our traditional ideas about diagnosis, prognosis and treatment. (See Tables 1 and 2.)

The above considerations open up several questions in the fields of psychiatry, psychology and psychotherapy: To what extent can certain psychiatric disturbances be classified as dynamical diseases? To what extent can certain healthy psychological phenomena be understood to be chaotic? To what extent can certain psychotherapeutic treatments be understood to be chaotic processes? We have attempted to answer these three questions in parallel from a semantic point of view onhand corresponding examples in another publication [21] already mentioned above.

There I have proposed that the behavior of any chaotic system (in both state- and real-space) displays the following

2.2.2 Six formal logical characteristics of chaos

whereby the above-mentioned general and special attributes are combined:

(1)   Dynamical self-organisation
           (=» long range correlations, "autopoeisis")
(2)   Intrinsic coherence with unpredictability
           (=» algorithmic complexity, "synchronicity")
(3)   Determinism with non-reproducibility
           (=» weak causality, "butterfly effect")
(4)   Complex recursive triadicity
           (=» complex self-reference, feedback loops, "self-awareness")
(5)   Reactive instability
           (=» bifurcations, lacunarity, "spontaneity", "lability")
(6)   Structural self-similarity
           (=» no characteristic time scale, fractals, "macro- in microcosmos")

Keywords which, under certain circumstances, may also be used to name each characteristic are presented in parentheses. An anthropomorphic translation of the above six characteristics would be: (1) "Chaos is creative." (2) "Chaos doesn't run, it dances with time." (3) "Chaos is extremely sensitive, but not random." (4) "Chaos is conscious of itself." (5) "Chaos is autonomous." (6) "Chaos is paradoxical / antinomic insofar as it contains itself within itself."

It may have already become clear to the reader, that very many well-known dynamical processes - whether they be from the fields of biology, medicine, physics, sociology, economics or elsewhere - can be described with these same six characteristics, so ubiquitous is chaos [12].

To avoid confusion throughout the rest of this work, I suggest that we call any system a "chaotic system" whose behavior *can* manifest itself as a chaotic process.

Consider now the following

2.2.3 Working definition of psychosis.

PSYCHOSIS: *An unusual phase of mental / emotional attitude ("mind/brain state") dominated by altered inner and outer experiences of such intensive subjectivity that these experiences as well as the behavior associated with them cannot or can hardly be comprehended by others and, consequently, banish the afflicted into isolation and lonliness. According to the International Classification of Diseases (ICD-10 [5]), a psychotic disturbance is manifest as long as at least one of the following symptoms is present: disorganized behavior, hallucinations, delusions, confused thinking.*

(As a consequence of this unusual mental state: Things are realized by the afflicted individual which are understood differently or not at all by others; The world is so pictured, as if it were being personally arranged for the afflicted or as if he or she were being manipulated by it; Beliefs are stubbornly adhered to and convictions are expressed which can hardly be brought into reasonable association with collective reality. For example, he or she may often feel as if they were in a hypnotic trance; he or she may hear voices; he or she may feel persecuted; he or she may believe that their very thoughts are being implanted, read, controlled or taken away.)

We can best compare psychosis with <u>dreaming:</u> Dreams also offer an alternative picture of the world (more precisely, a picture of our unconscious attitude) with a unique and often highly differentiated logic. Upon awakening or after remission of psychosis, one attains a certain distance to one's experiences. These can nevertheless leave a lasting impression upon one's future life. In any case, the "awakening" from a psychosis - not unlike waking up from a terrifying nightmare - is usually quite upsetting.

Let me now offer the following

2.2.4 Six formal logical characteristics of psychosis.

PSYCHOSIS: *A mode of behavior between the poles of disorganisation and compulsion. The COURSE OF DISORDER (but not necessarily the disorder itself - see below) manifests the same 6 formal logical characteristics as those of a mathematical-physical chaotic system, for example, the course of turbulent flow within a river, and accordingly displays "non trivial" (so-called "long range") correlations which only become evident between two or more modes of behavior over an extended period of time without possessing any particular, characteristic time scale.. The six characteristics are:*[b]

(1) Outbreak / remission: One speaks here about the "endogenic origin" or "spontaneous healing" of psychosis. (Self-organisation, long-range correlations)

(2) Easy recognition but difficult to impossible prediction of course (*autismus*): Attitudes, comprehension, interpretations are adhered to which fit neither the situation nor the expectations of others (*hallucinations, delusions*). (Unpredictability)

(3) Extreme sensibility of course to influences from the inner and outer world: Psychotic individuals are very sensitive, quick to be disturbed or frightened, are *ambivalent*, unstable and *confused* in thinking and feeling and tend to overreact (*disorganized behavior*) with exaggerated restlessness or social withdrawal. (Nonreproducibility)

(4) Complex circular amplifications in course development (hopelessly entangled feedback loops within an exaggerated self-consciousness, so-called "viscious circles"): The confused expectant attention of the individual to his or her social surroundings, psychic inner world and somatic constitution during the psychotic state induce intertwining "viscious circles" linking psychotic and nonpsychotic episodes. As a consequence, the afflicted person experiences unclear, amorphic borders between him- or herself and his or her inner and outer world (*Ego-disturbances*).

---

[b] Note: Certain psychopathological keywords which can be associated with each characteristic are presented in italics between parentheses within the following text. In addition, the respective formal logical characteristic of mathematical-physical chaotic systems (recall above) are also presented within parentheses after the text explaining each item.

Thus, the psychotic individual lives with a distorted self- and world-picture. This makes it difficult if not impossible for others to establish an empathic, accepting or intuitive contact with the afflicted person as well as for him or her to maintain a non-disturbed relationship with his or her own body, inner and outer world (*delusion*). (Complex recursive triadicity)

(5)   Sudden and unexpected, inhibited or free-associated thinking / feeling / behaving: The psychotic individual can't "make up his mind", hops to and fro from one state of thinking / feeling / behaving to another as well as between psychotic ("crazy") and nonpsychotic ("healthy") modes of comprehension and behavior (*ambivalence, confusion, disorganized behavior*). (Instability)

(6)   The variations in intensity of psychotic symptoms (state variables) have no characteristic time scale, i.e., they are similar on each of three very different time scales: the sociodynamic, the psychodynamic and the biodynamic. The processes on all three scales are obviously self-similar as well as also being similar to each other. In other words:

(i) The course of psychosis over a period of decades, years, months or days (sociodynamic time scale) determined, e.g., onhand medical records of hospitalisations and work, has an underlying dynamics similar to that of

(ii) the course over shorter psychopathological exacerbations lasting months, weeks, days, hours, minutes or seconds (psychodynamic time scale), e.g., onhand psychological and psychiatric third-person (AMDP, video etc.) or self evaluations within a single hospitalisation and this, in turn, also has an underlying dynamics similar to that of

(iii) the course of biopathological outbreaks within periods of hours, minutes, seconds or milliseconds (biodynamic time scale), e.g. measured onhand skin resistance, eye movement, EMG or EEG, within one single psychopathological exacerbation.

This may well be based upon projections of inner experiences into the outer world (transference) or introjections of outer events into the inner world (concrete, operational thinking and feeling). (Self-similarity: "Macro- in Microcosmos")

At this point, it is helpful to introduce a new important concept from somatic medicine.

The human organism is - not unlike an orchestra - an interplay of many different harmonies: dreaming-nondreaming, sleeping-waking, digesting, oscillating hormone levels, breathing, heart beating, oscillating brain electrical potentials etc. In other words: The human organism is an expressely *inadequate*   random generator! On the other hand, the human organism does indeed evidence many (chaotic) irregularities in its cyclic behavior. Such considerations bring us to the concept of

2.2.5 Dynamical disease

DYNAMICAL DISEASE: "The human body is a complex, spatially and temporally organized unit. In many different diseases, the normal organisation breaks down and is replaced by some abnormal dynamics. We have proposed that these diseases, characterized by abnormal temporal organisation, be called *dynamical diseases.*" ([7])

The *signature of a dynamical disease* is given by a marked change in the dynamics of some variable (e.g.. blood pressure, blood cell count, body temperature, pulse rate etc.). In this regard there are four basic types of qualitative changes:
- constant parameters develop large-amplitude oscillations
- new periodicities arise in an already periodic process
- rhythmic processes disappear and are replaced
  - by relatively constant dynamics or
  - by aperiodic dynamics ("chaos")
- aperiodicities (chaos) develop into
  - constant
  - periodic or
  - other aperiodic dynamics.

Here are a few *examples of recognized dynamical diseases* (see also [13])*:*
Repetitively or periodically recurring fluctuations of

Medical disorders:	digestive problems, swelling of joints, outbreaks of fever, migraine headaches, pain, edema etc.
Circadian rhythms:	sleep disturbances, jet lag, etc.
Hormonal dynamics:	cortisone level fluctuations, dysmenoerrhea, etc
Blood cell dynamics:	periodic fluctuations of circulating blood cells, periodic haematopoiesis, chronic myelogenous lukemia, aplastic anemia, neutropenia, etc.
Respiratory probs.:	Cheyne-Stokes respiration (large variations in ventilatory amplitude), Biot breathing (pauses between full breaths), SIDS (Sudden-Infant-Death-Syndrome), etc.
Cardiac dynamics:	cardiac arrhythmia parasystole, AV heart

	blockage, atrial fibrillation, multifocal atrial tachycardia, frequent ventricular ectopy, etc.
Neuropsychological dynamics:	periodic neurotransmitter disorders, bipolar depression, periodic schizophrenia, periodic panic attacks, periodic recurring night terrors, periodic somnambulism, etc.

If we analyse the idea of dynamical disease, it is evident that *a dynamical disease is a disturbance of the organism due to*

(1) a change in the normal space-time, dynamical organisation of the organism's functions

and not from

(2) mechanical injury, poisoning, infection from a foreign agent (bacteria, virus, mikrophagus etc.) or genetic development.

In other words, *a dynamical disease is a functional disturbance of the organism due to the abnormal flow of information within the system: organism-environment.* Such an abnormality might arise "spontaneously" within the organism itself or be induced by an "infectious" flow of information into the organism from its surroundings. Thus we conclude that:

Dynamical disease is an information processing pathology

meaning that information alone can make ill as well as heal!

(plasticity of neural networks)

By these disturbances it generally turns out that either the *normal* or the *abnormal* space-time, dynamical organisation of the organism manifests the six formal logical characteristics of chaos. In other words, either the normal or the abnormal flow of information within the afflicted system, organism-environment, is turbulent. Accordingly, the state space portait of this flow is in general fractal [15].

The possibility of such a "fractal structure of the psyche" motivates an understanding of certain psychiatric disturbances as dynamical diseases or *information processing pathologies.* In the realms of psychiatric / psychological / psychotherapeutic application, this interpretation has three decisive, practical implications.

• As just mentioned, certain psychiatric / psychological / psychotherapeutic disturbances can be understood as dynamical diseases or *information processing pathologies.*

• Regulation therapies, e.g., dietics, homeopathy, sleep-deprivation or light therapy and even psychopharmaceutical treatment can be viewed from a new, scientifically founded perspective. The effects of these types of therapy could lie in bringing the self-regulating human organism back to health by "juggling" his or her mental / physical condition from an unstable equilibrium state of illness to a state of well-being by means of a weak but exact "dosis" of just the right stimulus at just the right time ("chaos control", see below). The healing agent - the dynamical "juggling" intervention in the human regulatory system, - e.g. a diat, an herb, a scent, a musical rhythm, a light stimulus, a change in the sleeping / waking cycle, a massage, a sexual excitation, the stimulus of a so-called "mind machine" ([10], pp. 262-268) - is thereby an information carrier which must be applied within a narrowly defined range of intensity - not too little and not too much ("very low - but not too low! - doses and high dilution effects") - and only at the proper point in time - not too early and not too late.

• Clear directions for the therapeutic management of dynamical diseases or *information processing pathologies* can be made within the chaostheoretical framework. These primarily refer to three different strategies: (1) the amelioration of pathological chaotic developments by means of *nonlinear forecasting;* (2) the excitation of a jump away from a pathological dynamical development and toward a process of healing by means of *chaos control;* (3) the support of an existing tendency to self-healing and well-being, again by means of *chaos control.* (compare [26], p. 95).

2.2.6 *Consequences of the concept of dynamical disease to psychiatry*

Understanding psychiatric disturbances as dynamical diseases, i.e., as *information processing pathologies,* has several immediate implications (recall Tables 1 & 2): (1) certain *disturbances* may be indications of stubborn disharmonies between various dynamical "clocks" of the overall sociopsychobiological system; (2) *The topological characteristics* of certain attractors (e.g. subsets of points in the space of all possible emotional states) derived from mathematical time-series analyses of the above-mentioned disharmonies may help describe the dynamical characteristics of the sociological, psychological or neurological generators of the respective diseases; (3) *Illness* means an unstable homeostasis of the self-regulating human organism under the influence of an underlying attractor within a respective functional state space of symptom dimensions; (4) *Outbreak* of illness corresponds to a "jump away" from a "healthy" onto a "pathological" sociopsychobiodynamical attractor and can possibly be inhibited by the methods of "chaos control"; (5) *Diagnosis* corresponds to the identification of the corresponding pathological attractor, i.e., attractors are used as diagnostic markers; (6) *Prognosis* is dictated by the temporal stability of the "pathological" attractor whereby neither a perfection of the associated model of this attractor nor an improvement of the precision with which measurements are made in order to reconstruct this attractor can lead to reliable long-term predictions of the course of illness. At best, the methods of "nonlinear forecasting" [27] may allow short-range glances on the order of 1 to 5 time steps (defined by the time span between measurements) into the future; (7) *Remission* of illness corresponds to a "jump away" from a "pathological" and back onto a "healthy" sociopsychobiodynamical attractor and can possibly be induced by the methods of "chaos control".

Chaos research thus calls for an answer to the following question:

To what extent can certain psychological / psychiatric disturbances be understood to be dynamical diseases / information processing pathologies?

In our previous discussion above, we used the term "psychosis" in two conceptually different ways: On the one hand, in the neuropsychological sense of a particular *mind/brain state*; on the other hand, in the chaostheoretical sense of a *process* (time series of states) with an emphasis upon the course of illness as a historical sequence of psychotic and nonpsychotic episodes of consciousness. We concluded that the course of psychosis, i.e., the time series of outbursts of psychotic information processing in the brain, is chaotic. We did not yet, however, explicitly address the question of psychotic state as such, namely, that particular, unusual state of consciousness which we momentarily recognize in terms of the behavior of the afflicted individual and which we then interpret to be his (or her) momentary steady-state psychotic experience of the world and, consequently, call "his" (or "her") psychosis.

In the attempt to arrive at an answer to the question of chaotic state, we suggest the necessity of verifying or falsifying the following

2.2.7 Psychosis hypothesis

HYPOTHESIS: *The mind/brain state of psychosis is due to a decoupling or desynchronization of one or more (normally phased-locked) reciprocal connections within or between certain cell assemblies in the human brain.*

This hypothesis obviously stands in sharp contrast to our above understanding of the nature of (chaotic) psychotic course.

The decoupling or desynchronization referred to here results in a reduced nonlinearity in the information processing within or between certain cell assemblies in the brain. (A cell assembly is a distributed subset of neurons with strong reciprocal connections. A singular, specific cell assembly may underly the emergence and operation of every cognitive act. - private communication with Francisco Varela, Paris.) Accordingly, one may rephrase the above hypothesis by simply saying that:

The psychotic state is a linear information-processing pathology.

It would transcend the limits of this work to motivate and explain the ideas and observations leading up to the proposal of this hypothesis. Nevertheless, in order not to get confused between the concept of psychotic *state* itself as a linear process with a nonlinear, that is, chaotic, *course,* it is important to bear the following considerations in mind: Although the course of psychosis may indeed be chaotic, i.e. display long-range correlations between episodes and scalar invariance in the corresponding time series, the fragmentation of the psyche characteristic of psychosis itself (the mind/brain state, we actually recognize to be psychotic and, therefore, call "psychosis") may arise from linear information processing in the brain, i.e. may be *lacking* in long-range correlations between perceptions which enable the mind to perceive an overall "Gestalt". Indeed, a psychotic or pre-psychotic person is usually not able to understand his or her momentary situation meaningfully within its given context.

Linear information processing implies, for our purposes, that certain bodily-external and bodily-internal sensory signals superpose, one upon the other, resulting in a confusing blend of impressions unusual to our normal, nonlinear way of viewing the world. Such a blend would make it difficult, if not impossible, to cognitively and affectively differentiate between different simultaneous observations from external and/or internal signals. A linear state of mind/brain would thus be easily overwhelmed by too many parallel, that is, simultaneous, sensory signals and, thus, "not be able to see the forest for the trees", so to say. A brain functioning with certain decouplings between otherwise (nonlinearly) coupled cell assemblies would neither be able to filter out redundancies in incoming signals (from both inside and outside the body) nor would it be able to distinguish amongst the multiplicity of competing activation potentials which it is constantly forced to confront. Such a situation could, for example, lead to failure of the "Supervisory Attentional System" of the brain [23], [24]. This, in turn, would have several severe implications to the afflicted individual. For example, he or she, (1) might be incapable of voluntarily initiating (mentally activating) certain actions, (2) could be unable to supress one or another mode of behavior, and (3) could even be compelled (activated) to initiate certain other actions involuntarily, all of which might seem quite inappropriate within the context of his or her social setting: The people around him or her would experience this individual as being "psychotic"; he or she might experience themself as being "hypnotized", "controlled" or "manipulated" etc. and would, accordingly, offer his or her social environment a personal explanation of his or her subjective reality which, to third persons, would seem quite crazy.

Nonlinear information processing, e.g., hearing, can easily keep parallel track of many different simultaneous observations, e.g., several spoken conversations, all at once. Even though the time series of chaotic outbursts of psychotic linear information processing seem to manifest long-range correlations, each psychotic episode itself does not seem to enable the mind/brain of the afflicted individual to maintain long-range correlations of perception, i.e., does not seem to enable the mind/brain to adequately differentiate between simultaneous signal inputs and to embed these within a normal (nonpsychotic) cognitive/affective context. Thus, summing up, psychotic episodes of chaotically varying length embedded seemingly sporadically within an overall course of remission might correspond to outbreaks of linear (laminar) flow (psychotic state) occurring chaotically within a basically turbulent, long time process (state of remission). A good example displaying this kind of behavior is a waterfall whereby the normal flow is turbulent and chaotically occurring stretches of laminar flow (of chaotically varying length) are abnormal.

2.3 Second Aspect: Assessment of Chaos / Diagnosis of Illness

This aspect raises the question: *"How structured is chaos / psychosis?"* To help answer this question, we have to rely upon certain technical details (see e.g. [22]) which would transcend the scope of this paper. Expert readers will certainly be familiar with the terms I am about to introduce. For the lay person, however, I suggest a look into one or another popular introduction to chaos theory (also called "nonlinear dynamics") in order to fully grasp the meaning of this section (e.g. [8])[c] Nevertheless, if I have done my job well, even those not initiated into the deeper mathematical mysteries of chaos theory will be able to make a good overall picture of what I am trying to say.

From a technical point of view, chaos theory is concerned with reconstructing the geometry of an abstract state space from the time series of a single specific variable (see [17]). In the same spirit, process-oriented psychiatry is concerned with arriving at a diagnosis from the course of a single specific symptom. This stands in sharp contrast to classical psychiatry whereby the course of illness is presumed to be already contained within the diagnosis of the corresponding syndrome. In other words, classically speaking, the diagnosis implies all three: aetiology + course + treatment of illness. In a process-oriented psychiatry, however, the *form* of the course of illness determines both diagnosis and treatment. The aetiology of the disturbance plays here only a minor role. In the present-day, validated psychiatric classification schemes (DSM-IV [2] and ICD-10 [5]), the length but not the form of course plays the major role.

---

[c]Mathematically speaking, every point in state space corresponds to a single state and every state specifies certain fixed values for each of a complete set of independent variables (degrees of freedom). Translated for POPSY: Every point in psychopathological state space corresponds to a syndrome and every syndrome specifies fixed values for the severity (intensity and/or duration) of each of a complete set of independent symptom-features representing the axes or dimensions (degrees of freedom) of this state space.

The following table illustrates the correspondence between chaos theory and process-oriented psychiatry with regard to assessment of chaos / diagnosis of illness.

Chaos Theory	*POPSY*

Physical state space
State = all the independent variables (degrees of freedom) have a definite value
   = a single point in state space
	Psychopathological state space
	State = all the individual symptoms (degrees of freedom) have a certain severity
	   = a single point in state space

(a) Construction of the state space	(a) Definition of the state space
(b) Motion in state space (attractor)	(b) Course in state space (diagnosis)

<u>Ex.</u> Henon Attractor   <u>Ex.</u> Schizophrenic psychosis

Accordingly, POPSY research relies heavily upon observations of course during versus after psychosis onhand, for example,

(1) biological measurements such as electroencephalogram (EEG), electromyogram (EMG), skin resistance, eye movement, brain metabolism (positron emission tomography (PET)) etc.

(2) psychopathological measurements such as speech, behavior etc. from self- and third-person estimates

(3) sociological measurements such as those of social, living and work relationships, etc., for example, from a record of hospitalizations.

In this regard, the requirements of POPSY chaos research on the nature of collected data are twofold:

(a) With regard to the construction or the determination of the state space (Takens Embedding Theorem: [28]), we only need to observe *one, single* variable or symptom, but then

      (i) exactly enough (a scale of at least 5-bit resolution, i.e. 0 to $2^5 = 32$) and

      (ii) long enough (> 400 points);

(b) With regard to the motion or course in state space, we need a measure for chaos (= picture of the attractor = self-portrait of the "behavior generator"). Such measures are availabe from chaos theoretical analyses of the corresponding time series data and will be outlined below.

Now before looking at how chaotic time series data can be analyzed, let's review the

2.3.1   Three principally different classes of (non-quantum) motion.

These are illustrated below onhand mechanical, random and chaotic motion.

2.3.2   Mechanical (causal, linear, constant / oscillatory / periodic, "rational")

> *(e.g. Pendulum: Severity = angle from vertical; Var. 1=angle from vertical, Var. 2=angular velocity)*

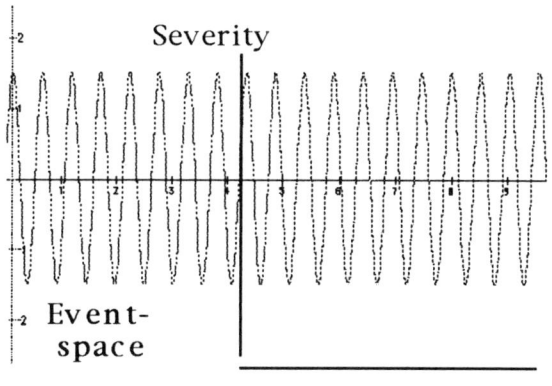

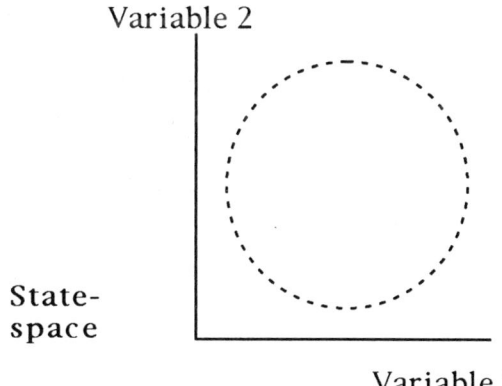

164

2.3.3 Random (acausal, stochastic, "irrational")

(e.g. Molecular motion of a gas in a bottle: Severity = x-position of a given molecule in the bottle; Variable 1 = x-position of this molecule; Variable 2 = x-component of momentum of this molecule)

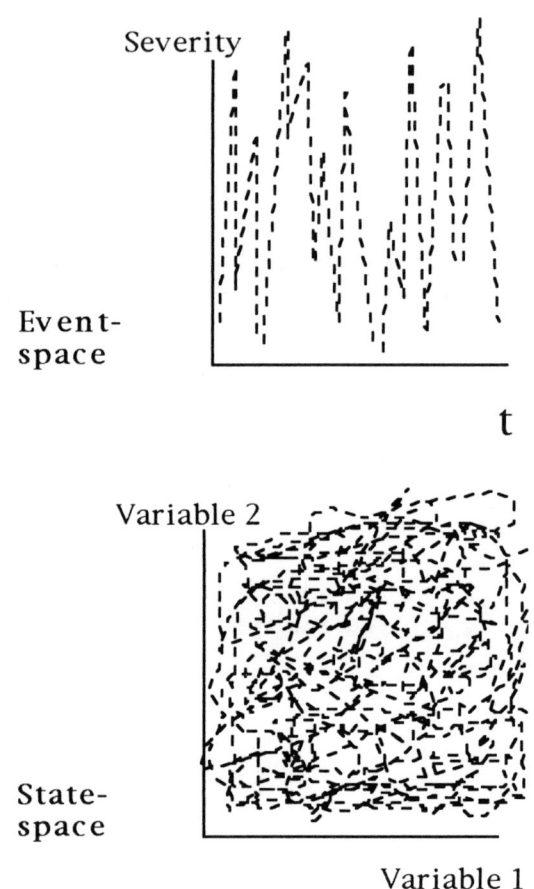

2.3.4   Chaotic (fractal, nonlinear, aperiodic, "disorganized" / "confused")

> *(e.g. Double pendulum: Severity = angle of "vertical pendulum arm" from the vertical; Variable 1 = angle of "vertical pendulum arm" from the vertical; Variable 2 = angular velocity)*

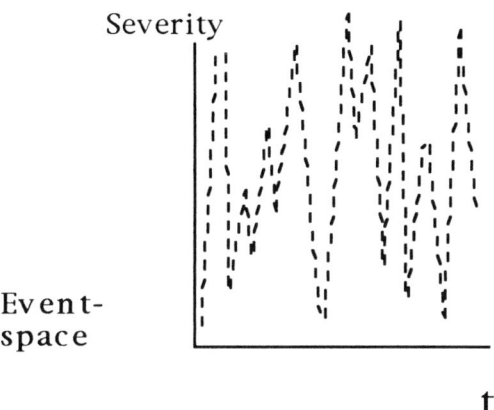

In the first case of the simple pendulum: The motion, i.e. every state of motion, is completely specified by the instantaneous angle (e.g. from the vertical) and the corresponding angular velocity of the pendulum arm. The magnitude of the angle is periodic with time and the corresponding plot in state space (angle versus angular velocity) displays at first a curved series of points which gradually closes upon itself to form a circle. Each point on the circle represents a physical state. As time goes on, the motion of points repeats itself over and over again along the same circular curve.

In the second case of a gas, let us imagine one particular gas molecule moving about in a closed room: The motion, i.e. every state of motion of this molecule, is completely described classically by its instantaneous position and corresponding velocity. The location of this molecule, e.g. from the middle of the room, is arbitrary with time and the plot of motion in state space (position versus velocity) shows nothing more than a cloud of points which gradually gets denser and denser as time goes on.

In the third case, we consider a so-called double-pendulum (=two pendulum arms hanging in tandem [20]): The motion, i.e. every state of motion, is completely determined by the instantaneous angles (e.g. from the vertical) and the corresponding angular velocities of both arms of the pendulum. The magnitude of the angle of the one or the other pendulum arm is apparently random with time and the corresponding plot in the state space of this arm (angle versus angular velocity) evidences at first a seemingly arbitrary clustering of points. As time goes on, however, this cluster gradually assumes a recognizable Gestalt which becomes ever more evident: Peculiar to this process is how the points skip about and seem to randomly fill out the figure with each successive appearance while not leaving the obvious bounds. In the case of the simple pendulum above, the points lie in temporal and positional succession, one after the other, along the circular curve and do not jump about from one to another position across chords.

In case of chaos it is possible, onhand the corresponding time-series data, to give a

2.3.5 Quantitative assessment of the 6 characteristics of chaotic (fractal) motion:

   (i) Degree of self-organisation
            - Correlation Dimension (D_2)
            - Kolmogorov Entropy (K_2)
            - $f(\alpha)$ Spectrum
   (ii) Degree of unpredictability
            - "Algorithmic Complexity"
            Reliable, individual forecasting:
                     » causal motion is in principle 100% predictable.
                     » random motion is only statistically predictable
                     » chaotic motion is predictable at the individual level only
                        over a very short time span
   (iii) Degree of non-reproducibility
            - Lyapunov Exponent (λ)
            (A measure of the exponential divergence of two initially
            neighboring trajectories)
   (iv) Degree of complexity
            - Correlation Dimension (D_2)
            - Kolmogorov Entropy (K_2)
            - $f(a)$ Spectrum
            - Algorithmic Complexity
   (v) Degree of instability
            - Lacunarity ("In<u>tra</u>-attractor chaoticity")
                     = unstable motion ("jumping") within one single attractor
                     (The system can even be hermetically sealed.)
            - Instability ("In<u>ter</u>-attractor chaoticity")
                     = unstable motion ("jumping") from one attractor to
                     another (The system interacts with the environment.)
   (vi) Degree of self-similarity
            - fractal portrait in state space (scaling exponent).

2.4 Third Aspect: Prediction of Chaos / Prognosis of Illness

This aspect raises the question: *"How coherent is chaos / psychosis?"* There are three principally different approaches to the answer to this question, each of which corresponds to one of the 3 different classes of (non-quantum) motion mentioned above:

- Mechanical motion	⇒»	Linear-reductionistic approach
- Random motion	⇒»	Statistical approach
- Chaotic motion	⇒»	Nonlinear-fractal approach

2.4.1 *Mechanical motion =» Linear-reductionistic approach*

This is a formalisation of the way we use our *causal*[d] *thinking* to solve practical problems:

> From the *last* of a series of states and the *time course* of corresponding sequential events, e.g., from the instantaneous location and momentum of a rocket as well as its previous trajectory, causal thinking allows us to *reductionally*[e] predict both the near and far future, in this example, the next points of flight up to the final location of impact.

The following diagram illustrates this approach.

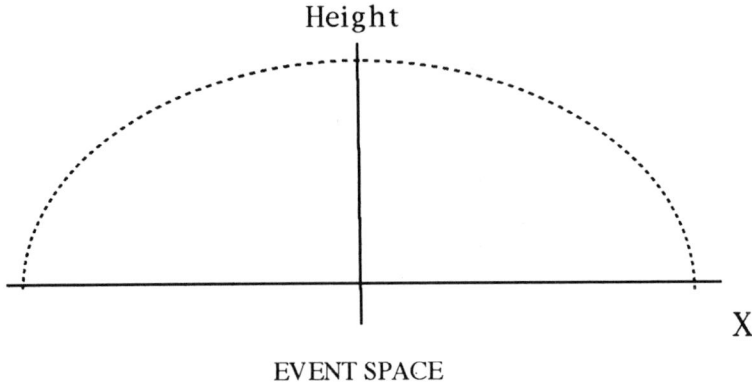

Height

X

EVENT SPACE

---

[d]"domino-like"
[e]"building-block-like"

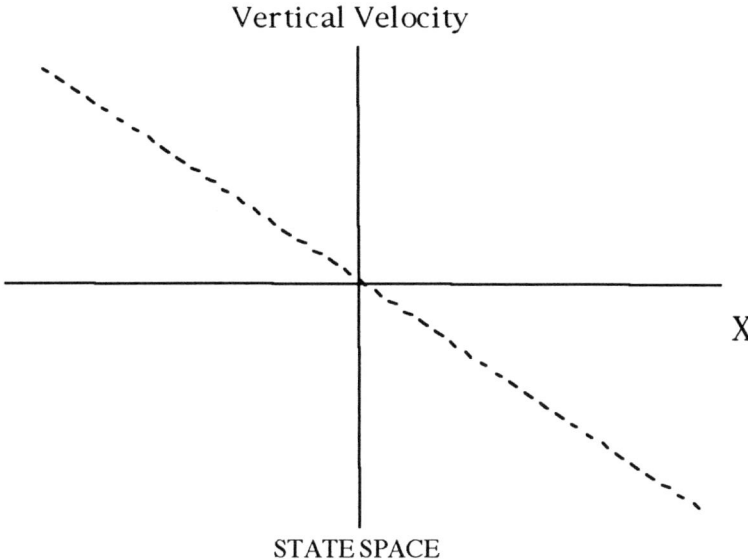

Vertical Velocity

X

STATE SPACE

2.4.2   Random motion =» Statistical approach

This is a formalisation of the way we use our *abstract thinking* in mathematics:

> From the given *distribution* of a *test sample* comprising a succession of previous events and the comparison of this distribution with a corresponding hypothetical basic population, the future, namely, the set of possible values for the state variables of the next upcoming event, is predicted with a certain *probability*.

A corresponding diagram to illustrate this point would show a familiar "bell curve" for the distribution of the test sample. (The "event space" diagram shows a random squiggle with time and the "state space" diagram contains a homogeneous cloud of points as already illustrated above for random motion.)

2.4.3 Chaotic motion =» Nonlinear-fractal approach

This is a formalisation of the way we use our *intuitive-associative* thinking ("common sense") in everyday life:

> Onhand the *common outcome* of a *set* of previous, *temporally independent events* (=points in state space) which are similar to the given event "here-and-now", we predict the immediate future *associatively*.

This formalisation is primarily made possible by the formal logical characteristics 1 & 6. (Divination, future telling, intuition etc. are included in this kind of thinking.)

The following diagram (a subset of the complete state space) illustrates this point.

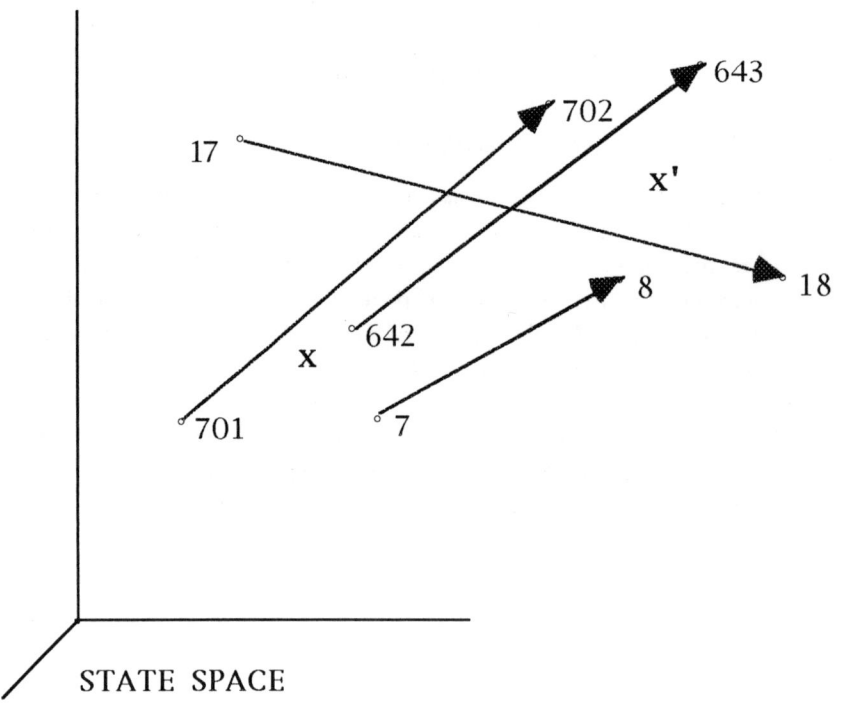

STATE SPACE

In order to predict the outcome of chaotic motion, one may employ the method of

2.4.4 Nonlinear forecasting":

> The "here-and-now" event point x is expected to arrive with the next time step (defined according to the measuring protocol) at the geometric center x' of that region of state space which has already been spanned by the positions of the immediate neighbors of x: 7, 17, 642 and 701 upon their past corresponding successive time steps to the respective points: 8, 18, 643 and 702.

(The "event space" diagram shows a seemingly random squiggle with time and the complete, corresponding "state space" diagram contains an elegant distribution of points as already illustrated above for chaotic motion.)

2.5 Fourth Aspect: Control of Chaos / Treatment of Illness

This aspect raises the question: *"How controllable / treatable is chaos / psychosis?"*

To help answer this question, consider a formalisation of the way we balance or juggle things. For example, imagine balancing a tennis ball on a horse saddle with tiny displacements of the saddle. We can either

(1) keep the ball in place on one and the same saddle or

(2) roll it over from the one to an other saddle.

In the mathematical translation, "tennis ball" stands for "the point in state space defining the momentary state of the system" and "saddle" stands for "attractor".

As a consequence, there are in principle

2.5.1 Two strategies for "healing" a dynamical disease / information processing pathology:

(1) In case a previously "healthy" attractor loses its attracting feature as a result of illness, one must constrain the system behavior to one and the same "nonpathological" attractor (in the spirit of "preventative medicine") or

(2) one must "juggle" the system behavior away from a "pathological" (e.g. a syndrome attractor in psychopathological state space - see above) and onto a "nonpathological" attractor by means of skilled therapeutic intervention.

In the first case, we cannot actually claim to be "healing" because we can never stop controlling the situation (in the sense of chaos control as mentioned above). In the second case, we do not have this problem and the corresponding treatment represents a rather general approach to the idea of chaos control.

The above overall formalisation is primarily made possible by the formal logical characteristics 4 & 5.

"Controlling" chaos (or a dynamical disease / information processing pathology) doesn't mean "quenching" it because the gentle adjustments of the system parameters leave the fundamental system dynamics - in our example, the shape of the saddle - unchanged. Quenching chaos (or a dynamical disease / information processing pathology) would, in our example, mean violently changing the shape of the saddle, say, into the shape of a wash basket (in order to make juggling unnecesary to keep contact with the ball)!

Tables 1, 2 and 3 present various perspectives on the differences between a traditional and a process-oriented approach to psychiatry.

3 IMPLICATIONS TO COMPLEMENTARY MEDICINE

The ideas presented above are not sufficient to simply go out and "heal" mental disturbances. Nevertheless I do believe that if medical researchers are ready to adopt one or another of the new concepts offered by the suggested chaos-paradigm, psychiatric research can take another step or two closer to solving some of the riddles, such as schizophrenia and psychosis, still facing us today.

The basic question for complementary medicine is "HOW do we *become* ill?" rather than "WHY *are* we ill?" Let us now take a look at the overall ideas presented above from each of these different standpoints.

HOW: The abnormal flow of information within the organism leads to a functional disturbance, a so-called "dynamical disease" or "information processing pathology" (see, for example, [1]; [7]).

Illness is regarded as a disturbance to the harmony in the space-time dynamical organisation within and around us ("sociopsychobiological clock"), so that thinking and feeling must first be directed into the proper channels and brought into "flow" in order for "healing" to begin. The question as to what or who had caused this flow to be pathological in the first place is not of central importance here.

The human sociopsychobiological clock is "correctly" adjusted by a gentle push of the pendulum - at the proper time (cf. the Greecian expression: "kairos") and in proper measure (very low - but not too low! - doses & high dilution effects). This "push" can be an appropriate audiovisual stimulation (with, for example, music, light or "mind machine" therapy), the right word ("AHA!-effect" during psychotherapy), or a significant life event (for example, falling in love, moving to a new apartment, accepting another job etc.). In this way, certain alternative and common sense healing factors gain a deeper interpretation.

WHY: Here we understand the ultimate cause of illness to be essentially biological: an injury, poisoning, a foreign agent of some kind (bacteria, virus, microphagus etc.) or a genetic development. This cause-and-effects-approach is in line with the classical concept of illness which is still dominant in the domain of mental health treatment today. Among the assumed (nonphysical) causes of mental illness, one also finds concepts like "double-bind communication", "not good enough mothering" etc. (see e.g. [4]) Although we actually have a pathological flow of information taking place here (e.g. between parent and child), that is, although the illness actually lies in system processes and not in the afflicted person themself (see below), the focus here remains the same, namely, upon *what* or *who* caused the illness.

The contrast HOW versus WHY emphasizes, in end effect, the contrast between "becoming ill" and "being ill". In the first case (becoming), there are no schizophrenic persons, i.e., the human being *is* not schizophrenic but, rather, certain social, psychological or biological circumstances *gave him* schizophrenia. The primary emphasis here is on prevention In the second case (being), he *is* a schizophrenic, i.e., he embodies schizophrenia or is identified with the illness schizophrenia. The primary emphasis here is on treatment.

In case the "sociopsychobiological clock" is not only out of synch but is actually physically defective, an answer to the question "Why am I ill?" could be "because of a broken wheel mechanism", whereas the question "How have I become ill?" would be more likely answered by "because the clock was thrown to the ground during a fit of anger." According to the first question, one searches for the *ill being* in the clockwork, e.g. in the schizophrenic person himself. According to the second question, one is not searching for the *being* of illness but, rather, for the process of *becoming ill*, e.g., in or during the rage of the agent who abused the clock. In the latter case, the illness is to be found in system processes and neither in the afflicted person nor in the agent themself.

4 SUMMARY & CONCLUSION

In the present work, we have explained the six formal logical characteristics of chaos from the standpoint of modern chaos theory: (1) Dynamical self-organisation; (2) Synchronicity; (3) Weak causality; (4) Complex self-reference; (5) Reactive instability; (6) Structural self-similarity (=Macrocosmos-in-Microcosmos). The relationships between these features and ideas / experiences from the field of psychiatry were discussed. In particular, these relationships led us to propose a new hypothesis concerning the nature and origin of psychosis (Point I below).

In conclusion then, let us review the four aspects relating chaos theory to a process-oriented psychiatry (POPSY).

I.   Certain psychiatric disturbances, e.g. the course of "psychosis",
        have six dynamical characteristics which can be closely related to the six formal logical characteristics of so-called "chaotic" processes. As a consequence, such disturbances can be understood to be "dynamical diseases" or "information processing pathologies".

        Hypothesis: Although the course of psychosis seems to be chaotic, the fragmentation of the psyche characteristic of the state of psychosis itself may arise from *linear* information processing in the brain (resulting from certain decouplings between particular cell assemblies which are normally interconnected in a complex, nonlinear fashion). As a consequence, parallel sensory signals superpose, one upon the other, leading to a confusing blend of qualitatively different impressions unusual to one's normal, nonlinear way of viewing the world which can easily keep parallel track of many different simultaneous observations all at once.

II.   *Chaos Theory:*   Geometry from the time series of a single
                          variable ("attractor in physical state space")

        POPSY: Diagnosis onhand the course of a single
                          symptom ("attractor in psychopathological state space")

III.   *Chaos Theory:*   "Nonlinear Forecasting"
Nonlinear forecasting enables the reliable, individual prediction of the behavior of a chaotic system for the next 1 to at most 5 time steps into the future. (A time step is defined in terms of the duration between measurements.)

POPSY: "Prognosis"
A reliable, individual prognosis is possible for the next 1 to at most 5 time steps into the future insofar as certain psychiatric disturbances can be understood to be "dynamical diseases" or "information processing pathologies".

IV.   *Chaos Theory:*   "Controlling chaos"
Chaos is to a certain extent controllable - comparable to controlling a juggling act or the growth of a plant.

POPSY: "Treatment of illness"
Certain traditional methods of healing as well as those known to complementary medicine receive a deeper, mathematical-physical interpretation.

5 CLOSING REMARK

Mythopoetically speaking:

A psychotic development is a spiritual crisis which *began* to *reflect* "too much" upon itself and thus to become "*self-similar*" in a *spontaneous* process of *self-organisation* with an *unpredictable and extremely sensitive* course violating its own sociopsychobiological limits and acting autonomously within the unconscious.

Such a crisis can shatter the soul of the afflicted individual into a thousand fragments. As long as just one sliver remains stuck in the mind's eye, this person can no longer view the world clearly, without distortion, and remain in normal contact with his or her surroundings. Such a state of mind is considered ill.

AKNOWLEDGEMENT

This work is an extension of a lecture presented at the international conference "Chaos, Fractals and Models" of the Society for Chaos Theory in Psychology and the Life Sciences at the University of Pavia, Italy, 25.-27. October 1996.

6   TABLE 1 » CLASSIFYING VS. PROCESS-ORIENTED PSYCHIATRY - GENERAL COMPARISON

Classifying Psychiatry	Process-Oriented PSYchiatry
- Disturbance / Illness	- abnormal space-time sociopsychobio-dynamical organisation
- Outbreak	- Attractor jump: "healthy" --» "pathological"
- Diagnosis	- Attractor = time-Gestalt of course
- Prognosis	- Stability of the attractor
- Remission	- Attractor jump: "pathological" --» "healthy"

7 TABLE **2** » CLASSIFYING VS. PROCESS-ORIENTED PSYCHIATRY - SPECIAL COMPARISON

Classifying Psychiatry: Emphasis upon …	Process-Oriented PSYchiatry: Emphasis upon …
- third-person organisation (order / singularity).	- self-organisation („chaos" / variability / multiplicity).
- diagnosis from cross-section (symptom profile).	- diagnosis from course (time-Gestalt of course: attractor).
- determining the diagnosis to allow for both long- and short-term prognoses.	- determining the attractor to allow for limited, short-term prognoses (at best).
- universal treatment of disease (time plan und impersonal dosis).	- individual treatment of disease ("kairos" und personal dosis).
- linear treatment.	- recursive ("internetted, systemic") treatment.
- external (biomedical) intervention in the healing process.	- self-treatment and autoimmune effects.
- gradual, induced changes.	- abrupt, spontaneous changes.
- a medically accepted concensus (canonical picture of health)	- a self-similarity between personal and third-person opinions (individual picture of health).
- "cementing" a fragmented psyche.	- "fractalising" a rigid psyche.

8 TABLE 3 » CLASSIFYING VS. PROCESS-ORIENTED PSYCHIATRY - THERAPY

Classifying Psychiatry	Process-Oriented PSYchiatry
Biodynamic Level	
- Suppression or elimination of the biological, chemical or mechanical cause of disease (Psychopharmaceuticals: Antidepressants, antipsychotic drugs, tranquilizers, seditives etc.)	- Support or reestablishment of the chrono*biological* regulatory systems of the organism. (Biorhythmic regulation: Audiovisual stimulation, light therapy, psychopharmaceuticals etc.)
Psychodynamic Level	
- Confrontation with pathological behavior patterns and pathogenic convictions. (Modification of disturbed conscious behavior: View of the patient from an objectively classifying perspective, e.g. via a cognitively oriented behavior therapy.)	- Support or reestablishment of the chrono*psychodynamic* regulatory systems of the organism. (Transformation of unconscious symbols: View of the patient through empathic identification, e.g. via a depth psychodynamically oriented psychotherapy.)
Sociodynamic Level	
- The individual is encouraged by his or her social environment to conform to the given social boundary conditions with minimum conflict. (Behavior therapy)	- The system "individual + social environment" adjusts itself to the given intra- and interindividual boundary conditions with minimum conflict. (System therapy)
Soziopsychobiodynamic Goal	
Function	Dialog

9 REFERENCES

1. an der Heiden U: Delays in physiological systems. *J Math Bio* **8**, 345-364 (1979).
2. Association AP: *Diagnostic and statistical manual of mental disorders. DSM-IV;* (4 ed. American Psychiatric Association, Washington D.C., 1994).
3. Benedetti G: *Todeslandschaften der Seele: Psychopathologie, Psychodynamik und Psychotherapie der Schizophrenie;* (4 ed. Vandenhoeck & Ruprecht, Göttingen, Zürich, 1994).
4. Chong DK, Smith-Chong JK: *Frag nicht warum ... Zur Struktur der Wirklichkeit und der Erweiterung unserer Fähigkeiten.* (Junfermann Verlag, Paderborn, 1995).
5. Dilling H, Mombour W, Schmidt MH, editors: *ICD-10: Internationale Klassifikation psychischer Störungen.* (10 ed. Verlag Hans Huber, Bern, Göttingen, Toronto, 1991).
6. Farmer JD, Sidorowick JJ: *Exploiting chaos to predict the future and reduce noise,* (Los Alamos National Laboratory, Theoretical Division, 1989).
7. Glass L, Mackey MC: *From Clocks to Chaos: The Rhythms of Life.* (Princeton University Press, Princeton, 1988).
8. Gleick J: *Chaos: Making a new science;* Cardinal ed. (Macdonald & Co. Ltd, London, 1987).
9. Goldberger AL, Rigney DR: Sudden death is not chaos in *The ubiquity of chaos,* ed. S. Krasner (American Association for the Advancement of Science, Washington, DC, 23-34, 1990).
10. Holler J: Das neue Gehirn: *Ganzheitliche Gehirnforschung und neue Medizin Theorien - Modelle - aktueller Forschungsstand.* (Verlag Bruno Martin GmbH, D-2121 Südergellersen, 1991).
11. Kloeden P, Deakin MAB, Tirkel AZ: A precise definition of chaos. *Nature* London **264**, 295 (1976).
12. Krasner S, editor: *The ubiquity of chaos.* (American Association for the Advancement of Science, Washington, DC, 1990).
13. Mackey MC, an der Heiden U: Dynamical Diseases and Bifurcations: Understanding Functional Disorders in Physiological Systems. *Funkt Biol Med* **1**, 156-164 (1982).
14. Mackey MC, Glass L: Oscillation and chaos in physiological control systems. *Science,* Wash **197**, 287-289 (1977).
15. Mandelbrot BB: *The Fractal Geometry of Nature.* (W.H. Freeman and Co., New York, 1982).

16. Morfill G, Scheingraber H: *Chaos ist überall … und es funktioniert.* (Verlag Ullstein, Frankfurt, 1991).

17. Packard NH, Crutchfield JP, Farmer JD, Shaw RS: Geometry from a time series. *Phys Rev Lett* **45**, 712 (1980).

18. Pool R: Is it healthy to be chaotic? *Science* **243**, 604-607 (3. February 1989).

19. Rapp PE, Latta RA, Mees AI: Parameter dependent transitions and the optimal control of dynamical disease. *Bull Mathl Bio* **50**, 227-253 (1988).

20. Rott N: A multiple pendulum for the demonstration of non-linear coupling. *J Appl Math Phys (ZAMP)* **21**, 570-582 (1970).

21. Schmid GB: Chaostheoretische Betrachtungen zu Psychiatrie, Psychologie und Psychotherapie: Teil 1: Die Sechs Grundeigenschaften des Chaos und eine Prozess-Orientierte Psychiatrie (POPSY) and Teil 2: Neue Hypothese zur Natur der Psychose. *Forschende Komplementärmedizin / Research in Complementary Medicine* (1997) **4**, 146-163 and **6**, * (in print 1997).

22. Schmid GB, Dünki RM: Indications of nonlinearity, intraindividual specificity and stability of human EEG. The unfolding dimension. *Physica D* **93**, 165-190 (1996).

23. Shallice T: *From Neuropsychology to Mental Structure.* (Cambridge University Press, Cambridge, 1988).

24. Shallice T: Can the neuropsychology case study approach be applied to schizophrenia? *Psychological Medicine* **21**, 661-673 (1991).

25. Skarda CA, Freeman WJ: How brains use chaos in order to make sense of the world. *Behavioral Brain Science* **10**, 161 (1987).

26. Spahr R. Chaostheorie und Psychologie: Ansätze zur Uebertragung (Lizentiatsarbeit an der Philosophischen Fakultät I, Abteilung Sozialpsychologie. Universität Zürich, 1990).

27. Sugihara G, May RM: Nonlinear forecasting as a way of distinguishing chaos from measurement error in time series. Nature, London **344**, 734-741 (19. April 1990).

28. Takens, F.: Detecting strange attractors in turbulence. in *Dynamical Systems and Turbulence* eds. Rand DA, Young L-S (Springer-Verlag, Berlin, Heidelberg, New York, Warwick, 1980).

SOCIAL ANTHROPOLOGICAL CONSIDERATIONS ON THE PREDICTABILITY AND UNPREDICTABILITY OF COMMUNITY OUTCOMES

GREGORY O. SMITH

Rome International University
Viale Dell'Università, 27
Rome, Italy 00185
E-mail: gosmith@iol.it

This chapter surveys community process in a circumscribed area of central Italy in a comparative effort to show how simple quantitative methods can provide insights into the nature of community constitution. It is evident that individual and psychological processes are rooted in community experience, and in order to have a fuller understanding of the various system levels discussed in this volume, it is valuable also to have some insights into the organizational dynamics of localized communities.

1 Introduction

The methodological tools associated with nonlinearity and chaos theory have, to my knowledge, never been applied to community studies. That such tools may provide fruitful insights in a sociological setting is a general idea I explored in a previous paper[1] which attempted to set out some of the methodological foundations for a reflexive self-organizing sociology where process yields outcomes which can only be understood as part of a nonlinear recursive event. The aim of that paper was to suggest an innovative resolution to the tension between a sociology focusing on individual agency, where choices are structured by an individual's changing perception of opportunity and expediency, and a sociology stressing the structuring role played by the overall system of relations. The interaction between these two levels of social experience can be characterized as recursive and nonlinear, and hence as falling within the purview of complexity studies. Borrowing broadly from Bourdieu's sociological approach[2] which looks at the a posteriori structuring effects of individual agency, I attempted to characterize the way in which the outcomes of individual actions condition the environment in which perceptions are formulated, establishing the assumptions upon which future choices are predicated. The complexity of this interaction, which is in a sense doubly hermeneutic, should be self-evident.

In the present paper I attempt to develop and expand this idea, sketching out the manner in which an accumulated set, or aggregate, of individual choices produces a collective outcome which structures the environment in which future individual choices are made, establishing a "path" having the force of what Durkheim[3] would call a "social fact". The observation that subjective creations have an objective quality has of course been common fare in sociology for some time (e.g., Berger and Luckmann[4]); what is innovative about the approach I am advocating is the assumption concerning the nonlinear character of recursive interaction within the system of social relations, an assumption which allows for a novel approach to correlating outcomes with different types of nonlinear paths. This is, be it clear, part of a larger project. In these pages I intend simply to show that aggregate outcomes can be measured using simple techniques of statistical analysis, and thus analyzed can help us infer something about the constitution of the community in which the specific outcome was attained.

The approach is broadly concerned with what Goldstein[5] calls the self-organizing features of social dynamics. It also bears affinity with reflexive sociology which places importance on individual perceptions, and game theory where individual choice is constrained by the types of knowledge available to the game players.[6] The specific application is to community studies, a branch of sociological enquiry which allows for comparative analysis which facilitates the combination of quantitative measures of community outcomes with a conventional anthropological, and hermeneutic, approach to community studies. This choice of a local, grassroots focus is also encouraged by the author's training as a social anthropologist. The following analysis concerns easily quantified economic outcomes measured among a small set of Italian townships.

2 The Setting

The towns selected for this analysis are located in Fucino, an excellent area for our analysis, since it is a homogeneous geographical unit containing ten different towns for which independent statistics of various types are maintained, including those regarding crop choice. The towns exhibit considerable individual variety in such things as dialect, local custom, certain geographical features of territory, and, importantly here, crop and animal choice. During the period considered, 1972-1982, Fucino was still heavily dependent on agriculture, and even today agriculture continues to play an important role in the local economies.

I first began my research in Fucino to investigate the sociological implications of its dramatic history. The area takes its name from a lake, formerly one of largest in Italy, which was twice drained, once in classical antiquity and once again in the nineteenth century. The conceptual implications of the drainage are intriguing, for an enterprise of such a scale required a vast injection of resources from outside the local arena, resources which were controlled by an external urban based power elite. The drainage thus brought to Fucino not only far-reaching geographical and meteorological changes, but also a new way of organizing economic and social relations. It would be easy to liken the event to a major catastrophe, in terms of the impact it had on the local life. Indeed, the transformation failed to bring promised prosperity, and Fucino remained poor and rather obscure; the wealth derived from the drainage flowed elsewhere.

The sociological implications of the drainage were complex. Here we have an area whose cultural way of life has deep historical roots, and then suddenly, in a period of forty or fifty years, the whole way of life was changed: a population of fishermen and shepherds was abruptly transformed into an agriculturally-oriented rural proletariat. The change, to borrow a metaphor used by Goldstein[7], was comparable to changes which occur when a liquid is exposed to a sharp increase in temperature. The system which over centuries had established a more or less steady state was broken, and a far-from-equilibrium condition was established in which the self-organizing features of the local communities, different as we might imagine such features to be in each town, had to come to terms with a single uniform set of constraints constituted by the new bureaucratic order imposed by the agents who had drained the lake, and by the new economic realities. There was also heavy immigration during and following the drainage - so the sociological implications were extreme.

The administration of the new lands by an absentee Roman landowner gave rise to an extraordinarily complex set of social relations which ultimately reached a state of system failure in the aftermath of the Second World War when the dynamic far-from-equilibrium condition which had been evolving over two or three generations finally collapsed. In the fifty or so years which followed the drainage a pattern of social relations did indeed emerge. The driving force in this pattern were the actions of a rural proletariat which has never been properly understood, since the farmers of Fucino, who were technically tenants, behaved as if they were smallholding agriculturalists, and, to complicated matters further, often day-laborers and shepherds in the bargain. The political rhetoric of the 1940s-50s fuelled a fascinating double hermeneutic process such that by the end of the 1950s no one had a clear idea of who the local people "really" were, including the local people themselves. Nor has history understood since. Fortunately, the present treatment requires only the briefest mention of these qualitative features, and focuses instead on quantifiable economic outcomes, which are fairly straightforward. Put simply, while the farmers of Fucino by the 1950s engaged in limited subsistence production, as they had done for millennia, they were essentially commercial farmers who sometimes relied on pastoralism as a sideline, but devoted most of their energies to the production of three main crops - potatoes, sugar beets and wheat. This pattern persisted in the period considered here.

While all three crops mentioned can be used to a greater or lesser extent to satisfy domestic needs (sugar beet pulp is sometimes fed to swine), the choice of how much to sow of any one crop is essentially a question of market rationality. The local farmers, who are technically peasants in the sense of being only imperfectly integrated in the national market, had full individual autonomy in making their economic choices in the historical period considered. Thus their choices were ostensibly conditioned by their individual and collective awareness of market conditions. Some markets were very stable. For instance, the sugar beet price is set by the EU before the agricultural season begins. The price of wheat is likewise known before farmers set their plows to the soil. Indeed, the only market which presents a high degree of uncertainty is that for potatoes, an uncertainty which is expressed both in terms of price and of how much produce the market can absorb. It should be added that the market conditions for these crops are identical for all the communities considered.

The hypothesis investigated here is that year to year variance in aggregate economic outcomes expressed at the level of the township will vary from town to town independently of objective market conditions. Indeed, in a perfect market situation we would not expect any change in variance from one town to the next, since the market constraints are uniform for all the towns considered. I further hypothesize that differences from town to town in yearly variance in aggregate crop choice - i.e., the predictability or unpredictability of community outcomes - will tell us something significant about the communities themselves. In order to explore these hypotheses I take the ten townships with land on the Fucino lake bed, and calculate, as stated, year to year variance. The broader aim of this analysis is of course to understand significant features of nonlinear recursive process at the local community level. But before turning to that analysis, let us see something about variety in variance at the town level.

3 The evidence

From the standpoint of our sociological model it is significant that even local discourse holds that order and disorder constitute significant variables in defining the character of local social relations. Educated people refer to peasants as an "amorphous mass" incapable of concerted action. Indeed, the educated take special pleasure in exaggerating the cultural poverty of the peasants they are associated with. In Celano, the town where I lived in the late 1970s, the local elite would state with satisfaction that "their" peasants were the most backward in the area. When discussing the problems which beset local agriculture, the educated observers claimed that farmers expressed economic choices with no rhyme or reason; "ignorance" and low social cohesion took the blame for economic failures. Even the local agricultural experts held that the chaotic nature of economic choices was a product of cultural insufficiencies, and the main cause of serious local crises. The impact these observations had on farmer choice is a relevant issue, but it is difficult to assess.

During my stay in Fucino in the 1970s my analysis of economic rationality in the area inevitably came to focus on the contrast between Celano, whose inhabitants had a consolidated reputation for being uncouth, and Luco dei Marsi, a town located to the west of the lake bed, where the farmers were considered to be advanced. The indices of advance were greater general prosperity, and a high degree of cohesion in community relations. For instance, Luco dei Marsi had many agricultural cooperatives, while other towns like Celano essentially had none, notwithstanding the formal requirement that farmers come together in cooperative associations. Perhaps because the cooperatives provided them with a greater degree of market security, the farmers of Luco dei Marsi invested heavily in expensive commercial crops like potatoes. Celanese farmers instead invested less, risked less, and diversified. Celanese farmers also often counted on low cost and generally small scale investments in sheep to practice, sometimes as a sideline, the solitary trade of the shepherd. If the farmers of Luco dei Marsi invested in livestock, it was generally a riskier variety which was maintained in stalls.

Undoubtedly some of the reasons for this difference can be traced to contrasts in the physical features of the two town territories. For instance, the Celano territory comprises a large mountain area which facilitates pastoral pursuits. Luco dei Marsi has no equivalent land area. Thus it seems natural that Celanese farmers should invest in sheep and goats. But physical differences alone cannot explain the full range of the economic contrast. For instance, it cannot explain why the people of Luco dei Marsi are prepared to take greater risks. Other factors must be called into play.

The Celanesi shy away from heavy investments in risky crops, and diversify their economic commitments in a variety of low cost ventures as part of a strategy in which pastoralism plays a significant role. Sheep raising may be a solitary pursuit which demands a large investment of labor, but it requires limited capital, and entails little or no risk. In Luco dei Marsi, instead, farmers concentrate on the highest risk crop, the potato, and on large livestock which also entails greater investment and greater risk than the small livestock kept in Celano. So clearly the perception of risk is differently framed in the two towns, representing a present embodiment of past experience which structures the future path of choice. Social constructs are involved here, not physical ones. It should be noted that the risk of potatoes is sizeable not only owing to the uncertainty of market, but also and especially owing to the high investment cost in terms of seed and fertilizer. A bad year for potatoes can be ruinous for the farmer who has overproduced. Yet the farmers of Luco dei Marsi take the risk, while in Celano the risk is accepted only on a much smaller scale.

When I carried out this study twenty years ago, lacking a more firm methodology on which to base my analysis, I followed the conclusion of the local elite, and argued that broad cultural variation determined the differences in economic orientations, suggesting that in Celano various features combined to make their agricultural choices more confused and chaotic than in towns like Luco dei Marsi. In the present analysis I attempt to enrich this hermeneutic understanding of local communities through the systematic quantification of aggregate community outcomes in such a way which modifies somewhat my hermeneutic understanding, while providing a deeper understanding of the linkage between individual perceptions and community structures.

The data on crop production for these ten towns over the period 1972-1982 lend themselves easily to quantitative analysis. Rather than analyze all three of the commercial crops produced on Fucino, which would require extensive treatment, for the purposes of the present exploration I have taken the crop for which market fluctuation is strongest, i.e., the potato, as the index of year to year variance, looking at the percentage of land on the lake bed within each town territory which was devoted to this crop each year. The differences from town to town are striking. The highest values for potatoes are those reached in Luco dei Marsi in 1977 when farmers planted potatoes on 64.92% of their arable. In contrast, Celanesi in this same year only grew potatoes on 48.63% of their arable. The lowest values for the ten-year period confirm the general trend: in 1981 potato production in Celano dropped to 22.6%, while in Luco dei Marsi potato production never dropped below 35.13%.

Here I must say something about how the data on crop production are classed. It will be seen in Table 1 that the Celano lake bed territory is divided between land which is contiguous with the town proper, and land which is located at some distance from the town, in the lowest part of lake bed, known as the Bacinetto ("little basin"). This territorial distinction is of considerable sociological relevance. Furthermore, the statistics for Celano include two neighboring towns, Cerchio and Aielli, all having territory divided between land close to the town proper and land in the Bacinetto. For the purposes of our analysis we shall consider three areas, what I call Celano proper, Celano Bacinetto, and Luco dei Marsi, all taken from the data shown in Table 1.

Table 1: Year to Year Crop Choice Variance for Fucino

Percentage Values by Surface Extension of Potatoes Planted (Source: Abruzzo Regional Reform Board)

Year	Avezzano	Celano, Cerchio, Aielli	Celano, Cerchio Aielli Bacinetto
1972	35.77	23.45	33.32
1973	43.67	31.64	28.3
1974	49.05	39.49	46.51
1975	43.64	30.65	31.94
1976	38.92	25.4	26.55
1977	59.91	50	48.63
1978	47.59	34.37	39.44
1979	44.29	35.19	36.45
1980	44.33	36.9	31.07
1981	37.12	26.06	22.6
1982	37.4	27.8	23.99
Average	43.79	32.81	33.53
Variance	47.59	28.81	73.51

Year	Pescina/S. Benedetto	Pescina/S.Benedetto Bacinetto	Ortucchio/Gioia
1972	29.08	33.66	22.87
1973	25.8	39.15	24.83
1974	33.27	38.12	28.45
1975	33.1	31.92	25.73
1976	51.52	26.54	42.17
1977	33.26	42.97	27.28
1978	39.49	32.2	36.58
1979	38.88	37.21	36.33
1980	29.3	36.22	34.64
1981	24.33	27.12	29.64
1982	27.45	22.73	24.93
Average	33.23	33.44	30.31
Variance	60.61	37.34	38.34

Year	Trasacco	Luco dei Marsi	Total
1972	26.34	37.52	29.88
1973	32.47	50.03	35.77
1974	50.29	55.37	43.09
1975	41.21	47.06	36.37
1976	40.7	46.72	33.51
1977	63.45	64.92	53.9
1978	44.26	45.73	40.32
1979	39.27	45.92	39.47
1980	47.33	42.06	39.2
1981	27.56	39.91	30.21
1982	31.43	35.13	30
Average	40.39	46.40	37.43
Variance	120.00	70.47	50.25

If we compare year to year variance in these three areas, we discover that the lowest value is that found in Celano proper, at 28.80. This is almost a third of the value found in Luco dei Marsi, at 70.47. Celano Bacinetto instead has a value comparable to the neighboring town, at 73.51. The crop trend in terms of the land area percentages devoted to potatoes is camparable between Celano proper and Celano Bacinetto, but contrast in variance values is strong. Celanese farmers when working in Celano seem to be less sensitive to annual variation in market trends, and vary their crop choice from year to year relatively little. One explanation for this may be that agricultural pursuits in Celano proper are treated as part of a domestic strategy in which conservative crop choices are matched to extensive use of pastoralism, especially during the winter months when herds cannot take advantage of mountain pasturage. Celano Bacinetto is more inaccessible to the town, and since there are few stalls or other farm buildings on the lake bed, it is less accessible to the marginal flocks which take advantage of the lake bed when it is not being used for crops. Since the Celano Bacinetto cannot satisfy as easily as Celano proper the multiple purposes of the low risk strategy, Celanese farmers seem to try and optimize market opportunities in the Bacinetto, attempting like the Luchesi to predict the oscillations in the year to year market trend for potatoes. However, true to their general economic orientation, they risk less on this risky market than do the Luchesi.

Clearly we are looking at a limited set of data. Other key considerations which can help us better to understand these communities would be the relative success farmers have in predicting the market, or the impact variation in potato production can have on other crops for which the market is stable. But even the limited data we have examined provide encouraging evidence that the method adopted here may be of utility. If we look farther afield than the two towns just examined, the results are even more striking. Consider the highest and lowest variance values for Fucino. Trasacco has the highest value in the area, at 120.00. Celano, already examined, has the lowest. The extremely high values in Trasacco might be related to the low degree of social cohesion which one ethnographer[8] found to characterize this town. It seems conceivable that in a more atomistic community competitive rivalries will be strong, and in a game of one-upmanship farmers may attempt to outmanoeuvre their colleagues by varying their annual crop choice in a striking fashion. However, we need more data and deeper analysis to support this conclusion.

4 Conclusion

While it is impossible at this stage to draw any firm conclusion, optimism regarding the sociological utility of the method proposed seems to be justified. Applied to an appropriate array of data concerning different areas of community performance, the method of measuring the predictability or unpredictability of community outcomes may provide an innovative way of understanding the communities we are studying, as well as something about social process in general.

I believe it can be fruitful to consider the parallels between the approach advocated there, and approaches implemented by other researchers who put individual agency and path dependence in the foreground. One example might be Putnam[9] in his analysis of regional variation in Italy. Putnam's findings, supported by an impressive array of quantitative data, indicate that cultural factors are the strongest predictor of institutional performance at the regional level. Working at the community level we may be able to explore allied issues even more deeply, to gain an understanding of how cultural factors operate while taking into consideration the dynamics of individual choice and perception, and aggregate community process. These dynamics involve a complex layering of perceptions and transactions. Even the perception of perception, that double hermeneutic process one finds in relations between farmers and elite, must have impact on how the community is experienced, and how individual choices are expressed. The channeling of choice involves a recursive process, in that choices are taken over and over again, outcomes established, and reiterative paths constructed which must give rise to patterns we can imagine, although never fully measure. We can however, measure a lower grade of system complexity, and that is the predictability and unpredictability of community outcomes, and attempt from there to say something more than can be conventionally said about how social systems operate.

References

1. G.O. Smith "Caos e fatalità" in Fra ordine e caos, eds. M.F. Turno, E. Liotta, F. Orsucci. (Edizioni Cosmopoli, Bologna, 1996).
2. P. Bourdieu Outline of a Theory of Practice (Cambridge University Press, Cambridge, 1972).
3. E. Durkheim Les règles de la méthode sociologique (Alcan, Paris, 1895).

4. P. Berger, and T. Luckmann The Social Construction of Reality   (Anchor Books, New York, 1967).
5. J. Goldstein The Unshackled Organization. Facing the Challenge of Unpredictability through Spontaneous Reorganization (Productivity Press, Portland, 1994).
6. B. Williams "Formal Structures and Social Realities", in Trust. Making and Breaking Cooperative Relations, ed. D. Gambetta, (Blackwell, Oxford, 1988).
7. J. Goldstein, Ibid., p.45.
8. C. White Patrons and Partisans. A Study of Politics in Two Southern Italian Comuni.(Cambridge University Press, Cambridge, 1980).
9. R.D. Putnam Making Democracy Work: Civic Traditions in Modern Italy (Princeton University Press, Princeton, 1993).

MODELS PORTABILITY: SOME CONSIDERATIONS ABOUT TRANSDISCIPLINARY APPROACHES

ALESSANDRO GIULIANI

Istituto Superiore di Sanita' , TCE lab.
Viale Regina Elena 299, 00161 (Roma) ITALY

FAX: +39-6-49902355
e-mail: MUT@ISS.IT

Some critical issues about the relative portability of models and solutions across disciplinary barriers are discussed. The risks linked to the use of models and theories coming from different disciplines are evidentiated with a particular emphasis on biology. A metaphorical use of conceptual tools coming from other fields is suggested, together with the unescapable need to judge about the relative merits of a model on the basis of the amount of facts relative to the particular domain of application it explains.
Some examples of metaphorical modeling coming from biochemistry and psychobiology are briefly discussed in order to clarify the above positions.

key words: mathematical modeling, complexity, multivariate analysis, soft sciences

1 Introduction

1.1 Physical explanations in non-physical sciences

The wide availability of powerful computers together with the development of non-linear dynamics made a lot of new fields amenable to quantitative modeling (11,38,39).

The science of complex systems is the natural "all-embracing" frame for the quantitative modeling of fields dealing with very complex problems like psychology or biology (3,7,15,39,44).

The analysis tools of complex systems science come from physics and take with them the physical concepts from which they originate. The dependence of data analysis tools from physical concepts and theories, obviously, is not a peculiar property of non linear science; the two more universally used statistics, i.e. the mean and the variance, derive from the physical concepts of barycentre and inertia momentum respectively (36). But, while in the case of the mean and the variance of a distribution, the physical derivation is so distant in time to have no more influence on the practical use of these statistics in other application domains, the use of the recently derived non linear descriptors is still strictly associated with the correspondent physical concept (12,36,40).

This fact constitutes a great temptation for the user of the non linear tools in sciences different from physics: the temptation to build a physical-inspired theory of his/her field of interest, given the applicability of the particular non linear descriptor he/she adopted for describing the data.

This is a very risky attitude, because physical concepts are not dogmatically applicable to soft sciences like economy, psychology or biology; neverthless the metaphorical use of physical theories to describe complex phenomena has a great heuristic power to hihlight different views of the studied systems (1,2,3,11).

In this work I will try to sketch some recommendations for using the heuristic power of physics-inspired modeling in soft sciences without falling in the traps of motiveless theorization.

As first I will briefly discuss the differences between soft and hard sciences having in mind biology and physics, then I will describe two examples of metaphorical use of physical concepts in biology at different definition levels.

1.2 Soft and Hard Sciences

Traditionally, sciences may be grouped into two categories: "hard" sciences, where repeatable events can be predicted from a mathematical formalism expressing laws of nature; and "soft" sciences of complex systems, where in many occasions only a narrative account of distinguishable events is possible.

While mathematical modeling of hard sciences, like physics, is generally based upon differential equations, this style of reasoning can hardly be applied to soft sciences like biology (11).

Modeling through functional laws, typical of physics, is based on a number of requirements that are seldom satisfied in biology. Particularly important requirements are (11):

1) previous identification of the objects that play a relevant role in the phenomenon under study

2) identification of the boundary conditions of the applicability domain of the model

3) clear indication of the scale (or scales) in time and space most relevant for the studied phenomenon.

The fulfilling of the above criteria makes it possible to correctly evaluate the validity domain of an empirical finding.

In order to relax the above constraints and then make the "soft" sciences amenable to a quantitative treatment, different strategies can be devised. The most commonly used approach consists in trying to build probabilistic models in which the analysis of the single atomism of the system is replaced by the study of time and/or ensemble averages over a huge number of events.

The probabilistic approach, reminiscent of the thermodynamics style of reasoning, allows to build models without having a thorough theoretical knowledge of the system at hand (7,13,15): we need only to fulfill the above described requirements at the distributional level. Obviously, the information we can derive from such kind of modeling only holds at the "coarse grain": we can not describe the system at the level of the single atomism (15,38).

There is a very big debate regarding the strategies to detect the level of "coarseness" able to maximize the information derivable about a particular system (20,26,38,41). Roughly speaking, the problem is to choose the scale of observation that maximizes the ratio of "deterministic" non-trivial information to stochastic fluctuations (41). In hard sciences this is not a problem, because it is assumed that the most fundamental level is the more informative one (15,39), in soft sciences it is impossible to automatically put the fundamental microscopic level as the most informative one (11,15,38,39,41).

New techniques and the availability of more powerful computers have led to the development of highly detailed models in which a wide variety of components and mechanisms can be incorporated. In a model of animal grouping, every animal can be tracked; in a forest model every tree; in an epidemiological model, every individual in a population (38). Because models of this sort may provide an unjustified sense of verisimilitude, it is important to recognize them for what they are: imitations of reality that represent at best individual realizations of complex processes in which stochasticity, contingency and nonlinearity underlie a huge multiplicity of possible outcomes (38). Individual simulations cannot be taken as more than representative of this diversity, but repeated simulations can provide statistical ensembles that contain robust kernels of truth (38).

The success of a particular modeling is crucially dependent from the possibility of identifying a global phenomenological level, which is largely autonomous from the particular microscopic level realizations, and is easily described by statistical formalisms (41). In order to predict a pathological status, we only need to know the degree of variability of heartbeat over thousands of beats and not the causes underlying the duration of each single beat.

The important thing to keep in mind is that in biology we deal with phenomenological, coarse-grain, regularities and not with theories able to deterministically connect different levels of organization. This is due to the crucial role played by individuality, i.e. to an irreducible level of variability not simply assimilable to noise (11,18,19,27,31,33,44).

The same system, observed at different scales, can be viewed as made of rule-obeying interacting components or driven by essentially stochastic effects. The calcium channels are a wonderful example of this duality (19): a single channel is viewed at a structural level as a simple ensemble of proteins of known sequence and structure (fully deterministic), neverthless its behavior is modeled by a Markovian stochastic sequence of low and high permeability states (19) and two structurally identical channels can exhibit a completely different dynamical behavior (the subsequent description of protein dynamics will justify this discrepancy) (19,25).

To this respect, a very interesting case-history is the progressive decline of the importance of pure deterministic chaos in the explanation of physiological systems. In the eighties many mathematically inclined physiologists claimed to have found evidences of deterministic chaos in a lot of physiological systems (5,16). This idea stemmed from the application of mathematical tools like Lyapunov exponents or correlation dimensions; the physiologists, given for granted the existence of an attractor-like behavior for physiological systems, simply transferred the physical concepts to the biological reality.

Progressively, became more and more apparent that physiological systems are not necessarily best thought of as being fully deterministic in design (17,19,29,46,47,48,51). Independent of the number of degrees of freedom present, completely deterministic systems are simply fixed and rigid whether they be periodic or chaotic (47). Such dynamical systems, relying for the adaptation to novelty on an immediate change of parameters in response to external stimuli, are quite unprepared to exist in the noisy environments typical of the real world (29,46,47).

This by no means implies that the idea of deterministic chaos is useless in biology, on the contrary this idea stimulated the search for meaningful dynamics (i.e. for detecting useful information beside the simple statistical, order invariant, level of knowledge) not simply linked to the deterministic chaos paradigm (51). In fact there is a growing body of evidence of an intermingled mix of stochastic and deterministic components in a lot of physiological systems (17,19,22,29,48).

From a practical point of view, this forces the investigators to look for "soft" dynamical analysis tools able to detect deterministic structures embedded into a stochastic environment like Singular Value Decomposition (SVD) or Recurrence Quantification Analysis (RQA) without making any assumption of an attractor-like behavior of the studied system (2,8,11,13,21,28,43,45,46).

2. Metaphorical Modeling

2.1 Ising Models and Spin Glasses

Diversity is the first prerequisite of complexity : a complex system should show many significantly different states (25). Another characteristic of complex systems is "contingency", that is to say, the dependence of the present state on the vagaries of its past history (23,25). In addition to these points we must mention the contemporary presence of different scales at which the system displays meaningful structure (10,11,15,38,41,49).

The most simple and general model showing all these features is the Ising model of spin glasses (11,27,37,42), this is the reason why this model was applied in a so large array of situations going from the study of superconductivity (14,15,49) to the research on neural networks (1,30,42). Before going ahead to analyze how this model was applied in a metaphorical way to very different biological situations I will briefly introduce the underlying fundamental concepts.

The Ising model derives its name from Ernst Ising who first introduced it in 1924 (37). The problem the model deals about is to try and explain the cooperative phenomena which characterize the order/disorder transitions: the context in which the model was introduced is the phenomenon of magnetic susceptibility but, in more recent years, the same model was found able to explain a huge class of critical phenomena like turbulent flow (15,37,42,49), the onset of superconductivity (15,49), the conformation of polymers (22,23,24,32) and so forth (4,35).

The ferromagnetic materials ,like iron, display spontaneous magnetization; on a microscopic level this means that the spins of the constituent atoms have a preferred orientation. At high temperatures iron does not show spontaneous magnetization, when cooled at a very specific temperature called Curie point, the magnetization suddenly appears as a consequence of a critical degree of ordering of the magnetic moments of the constituent atoms (15,37,49).

This ordering has to be intended in statistical terms: each single spin tends to orient itself parallel to its neighbors but this tendency is contrasted by thermal agitation which randomizes the relative orientations of the moments. The struggle between these two forces is the responsible of the critical behavior we have described (14). This critical behavior can be explained with a very simple schema of a regular bidimensional lattice in which each microscopic magnet (atom) "sees" four neighbors (Fig.1): given a certain probability for two neighbors to be parallel, the curve describing the specific heat of the system varying the temperature displays a singularity correspondent to the

Curie point (37,49).

Fig.1

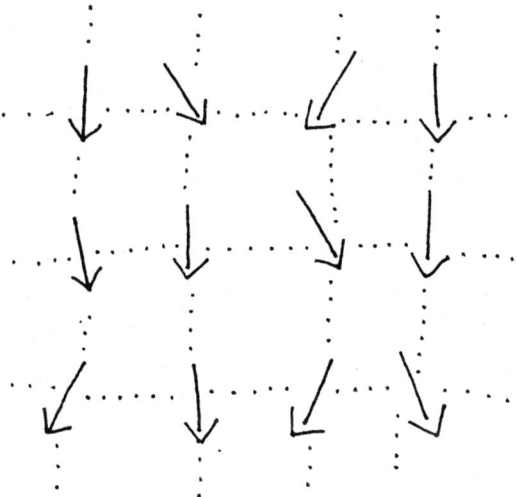

For our purposes, the important point is that the magnetization (and thence order) is a gross scale feature, the material, observed at the microscopic scale, appears randomic and completely dominated by the thermal agitation. Kenneth Wilson, in the seventies, gave a very elegant demonstration of the onset of critical phenomena in this kind of models using the so called "renormalization group" approach (15,49).

This approach is based on the computation of the average spin at subsequent scales (blocks of n spins, blocks of n blocks, blocks of n blocks of n blocks,....), at each step the variability of the precedent level is eliminated and only the average value survives, so obtaining separate descriptions of the system for each scale (27,49). The simulations of bidimensional Ising lattices analyzed in this way, displayed the expected critical behavior; for this work Wilson won the Nobel prize in physics.

In order to fully understand the metaphorical use of this model in biology we need to explain two more points: the behavior of paramagnetic materials and the concept of frustration.

Ferromagnetism is an intrinsic property of the atoms, this is the reason why, below the Curie point, iron shows spontaneous magnetization; on the contrary there exist some substances called paramagnetic, displaying magnetization only when embedded into a force field which acts as an external order parameter orienting the spins (6,9,14,33,42,49).

The case of frustration has to do with the so called glassy behavior: let us assume that in a given material we have two molecular species, one (A) displaying ferromagnetic behavior, i.e the tendency to assume a parallel orientation for neighboring spins, and another (B) with an antiferromagnetic behavior, i.e. the tendency to assume an antiparallel (opposite) orientation for adjacent spins. Each atom of such a material "sees" both A and B neighbors and cannot contemporary fulfill all the competitive constraints, this situation, using a psychological analogy, is called "frustration" (42). Such a material has a lot of "energy minima", i.e of possible arrangements (structures) which can minimize frustration without fully eliminating it (1,25,32,42).

These materials are called "spin glasses" because, in analogy with glasses, do not have a unique minimal energy crystalline conformation. The spin glasses can be considered the simplest systems displaying the hallmarks of complexity we sketched above (1,25,42).

In fact the spin glasses display diversity in terms of different minimal energy states, they show contingency, because the particular visited minimum depends on initial conditions and on environmental inputs and finally they display structure at different scales (1,42).

The Ising models are very general and simple to use: they have very few ingredients and could be simply translated in general statistical terms substituting positive and negative correlations for the ferromagnetic and antiferromagnetic interactions respectively and error variance for temperature (1,27,30,33,35).

2.2 Spin Glasses and Proteins

In textbooks, biomolecules are usually depicted as having a well-defined structure, and the word "dynamics" normally does not appear in the index (23). However, even for the simplest function, such as the entrance of dioxygen into myoglobin, the protein must fluctuate and fluctuations or motions imply a multitude of structures or conformational states (CS) (23,24,25,50).

At any given instant, an individual protein molecule is in a specific CS (24). It usually does not stay there indefinitely, unless at very cool temperatures, but moves through different CS exploring its energy landscapes (Fig.2). The CS can be equated to the "local minima" of spin glasses, and constitute a "rugged energy landscape" which the molecule incessantly explores (23,50).

It is useful to contrast two types of motion in proteins: relaxation processes and equilibrium fluctuations (24). In a relaxation process, the system moves from a non-equilibrium state toward equilibrium. The non-equilibrium state can be created by a chemical reaction or a temperature or pressure jump and corresponds to a peak in Fig.2 (24). The equilibrium fluctuations occur even in quiescent molecules and lead from one CS to another (valleys of Fig.2) (24). It is possible to arrange the energy landscape of proteins in a hierarchical way, the hierarchy coming from the different heights of the energy barriers between states (Fig.2), this hierarchical organization is at the basis of the presence of structure at different scales (23,24,25,50).

Fig.2

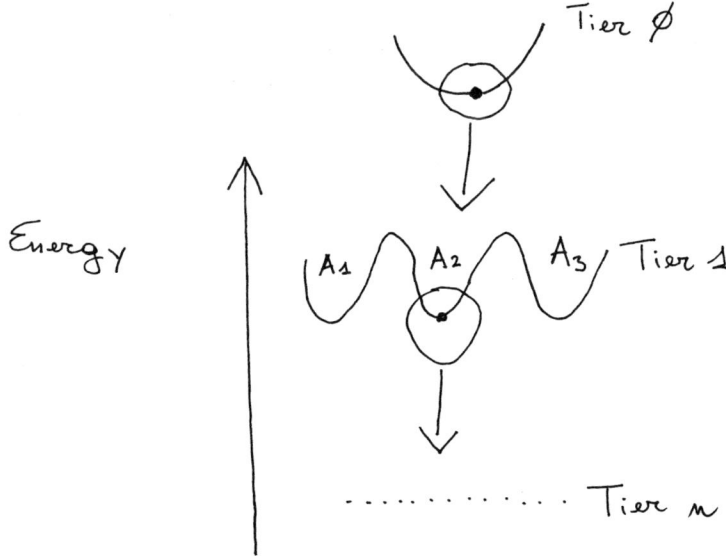

Each level of structuring corresponds to different aspects of protein physiology: folding, enzymatic catalysis, allosteric regulation and so forth (50).

The general resemblance (metaphor) between proteins and spin glasses gave the idea to Hans Frauenfelder to try to apply the phenomenological law describing the fluctuations in glasses to protein motion (25). In glasses the time dependence of these fluctuations is described by the subsequent formula (24):

$$F(t,T) = \exp(-(k(T)t)^b) \qquad (1)$$

In (1) the F is related to the variation of some observable related to the structure of the system at hand (in proteins is the position or the intensity of a suitably chosen spectral line), t is time, T the absolute temperature and k and b are experimentally derived parameters describing the system at hand.

The general model described in (1) is denominated "stretched" exponential (24,32) (in opposition to the classical exponential relaxation processes governed by Arrhenius law) and accounts for the fact that the relaxation function extends over many orders of magnitude in time (multiplicity of scales) (24,32).

In glasses the application of (1) gives rise to two kinds of motions: large scale (relaxation processes) and short scale (equilibrium fluctuations) so offering another point of similitude between glasses and proteins (24,25,32).

The (1) was discovered to be fully applicable to proteins and the qualitative features of protein dynamics were found to be completely superimposable to the spin glasses ones (23,24,25).

Why the phenomenological law of spin glasses applies to protein motions ? What proteins have in common with spin glasses ?

The resemblance between these two systems is based on the concept of frustration: frustration arises in folded proteins, for instance, because side chains have to compete for positions that minimize their relative energies. Imagine a polymer with pairwise short range interactions between the amino acids (the basic constituents of proteins) that may be attractive or repulsive. They might, for example, be hydrophobic or hydrophilic (25). When two segments of a string of such residues in an heteropolymer are brought together by folding, there will be both positive and negative contributions to the configuration energy. It is very hard, if not impossible, to organize the string locally so that the various phobias (and philias) are all accommodated (25), then frustration occurs together with the consequent multiplicity of local energy minima. The basis of the metaphor is thence situated at a very deep structural level and this gives to the metaphorical modeling a great explanatory power (i.e. it accounts for a lot of features of protein dynamics). It is important to keep in mind that, notwithstanding that the proteins ARE NOT spin glasses, the knowledge of spin glasses laws gave the scientists a conceptual frame to describe proteins motions.

The other example we will discuss deals with a much weaker resemblance between spin glasses and the modeled biological system, neverthless in this case too the metaphor can be established so giving us some useful hints about the studied system.

2.3 Spin Glasses and Pavlovian Conditioning

This paragraph deals with the application of ideas and data analysis tools coming from the study of Ising models of magnetization to the dynamics of learning in rodents (11,27).

The experimental paradigm analyzed was the so called Active Avoidance using a shuttle box. A shuttle box is a cage divided into two compartments, one compartment is electrified thus causing a shock to the animal inserted in it (27). This negative stimulus is preceded by a flashing light, thus teaching the animal to associate the flashing light with the shock (11,27). After a certain period of training, the animal learns to avoid the shock escaping to the non electrified compartment upon the presentation of light. The animal's performance is measured by the time (escape latency) it takes to leave the electrified compartment upon seeing the flashing light. Each training trial consists of the presentation of a light-shock pair, the entire training consists of 6 daily sessions of 40 trials each (27).

The goal of the investigation was to elucidate the temporal scaling of learning process, i.e. to discriminate between the hypothesis of a continuous learning dynamics in time (existence of a single characteristic scale of learning) and the hypothesis of different independent time scales at which learning takes place (multiple scales of learning correspondent to different neurological processes).

The experiment was executed on two rodent species (mice and rats) so to obtain a greater generality, here we will report the data relative to mice but the results were totally coincident for the two species.

The metaphorical linkage between this problem and the Ising models was established considering learning as an order parameter orienting the performance of animals in time progressively reducing their escape latency (27,34). The escape latency at each trial for each animal plays the role of a single atom of a paramagnetic material, the degree of order imposed by the order parameter is measured in terms of correlation coefficient between escape latency and the correspondent time measure at the different scales (trials, blocks of trials, sessions) (27,34).

The renormalization group approach was applied to the data (on an animal by animal basis) in order to compute these correlations at the different time scales. The computation of such correlations was executed at the single trial scale (correlation measure = CORRSS), at the session scale (correlation measure = CORRLS) and at the 4, 5, 7, 8, 10 trial blocks scales (correlation measures = CORM#S) (27).

After the computation of these correlations, the data set can be thought as a multivariate matrix having as rows the single animals and as columns the different correlation measures. In the hypothesis of a single common learning characteristic scale all these correlation measures had to be related each other. A principal component analysis (PCA) applied to the multivariate matrix highlighted the existence of two independent scales of learning (components) correspondent to two independent learning modes: a mode operating on the short time scales of trials and blocks of trials and a large scale mode operating at the level of daily sessions. This duality is evidentiated in Table 1 reporting the factor matrix (8,36) relative to the PCA. The rows of Table 1 are the correlation coefficients between time and performance at the different scales (original variables), the columns represent the relations (ranging between 0 and 1) of the original variables with the extracted components. As can be seen from the table a two components solution explains almost all the data set variability pointing to a long range ordering (PC2) linked to the sessions scale and a short range ordering (PC1) linked to the trials scale. These two orderings are independent by construction (8,11,12,27,36).

Table 1: Factor loadings matrix and percentage of explained variability

Variables	Components	
	PC1	PC2
CORRLS	0.02	0.99
CORRSS	0.98	0.05
CORM4S	0.91	0.12
CORM5S	0.97	0.08
CORM7S	0.98	0.07
CORM8S	0.99	0.03
CORM10S	0.96	0.04
% expl. var.	80.1	18.3

The existence of two independent orderings at the different scales, forces us to imagine different neurological bases for the small and large scale learning and an "inter sessions" consolidation period consistent with the existence of an internal representation of the task "reverberating" in the animals mind (1).
In this example the linkage of the biological phenomenon with the Ising models is not so deep like in the case of proteins, being based on the simple consideration of learning as an order parameter; neverthless we can make use of data analysis tools developed for the Ising models (the renormalization group approach) which allow us to develop a temporal scaling of the system qualitatively resembling the spin glasses (presence of structure at different scales).
In both the analyzed examples, the relevance of the model resides in the explanation given to experimental facts pertaining to the peculiar domain of application, i.e. biochemistry and psychobiology and not in the simple demonstration of the applicability of the Ising models. The proteins work derives its relevance from the fitting of the theoretical parameters derived from spin glasses with experimental observable of protein dynamics (32), the psychobiological work derives its validity from the independent demonstration of classical features of animal learning (34) with the added value to give the possibility to candidate otherwise purely qualitative observations.

2.4 Fractal...so what ?

On the contrary, a lot of applications of non linear dynamics concepts, seem to be
motivated by the only aim of demonstrating that something can be described by a given
mathematical formalism. A classical example is the plethora of works claiming for a
"fractal nature" of objects and processes.

A system having a "fractal nature" is a system with one or more characteristics that
remain constant when examined over a wide range of scales (10). A fractal boundary,
for example, would be one that appears as equally invaginated or "crumpled"
regardless of the magnification with which it is examined (10). A characteristic of such
boundary is that its apparent length increases as shorter and shorter standards are used
to measure it. The change in apparent boundary length is related to the length of the
measuring stick in a deterministic way (10).

Mathematical fractal are generated by recursive expressions wherein each generation
is derived from the preceding in a specific way. The basic fractal expression is a
summarizing statement describing a recursion (10). In a single dimension the idea is
diagramed in Fig.3.

Fig.3

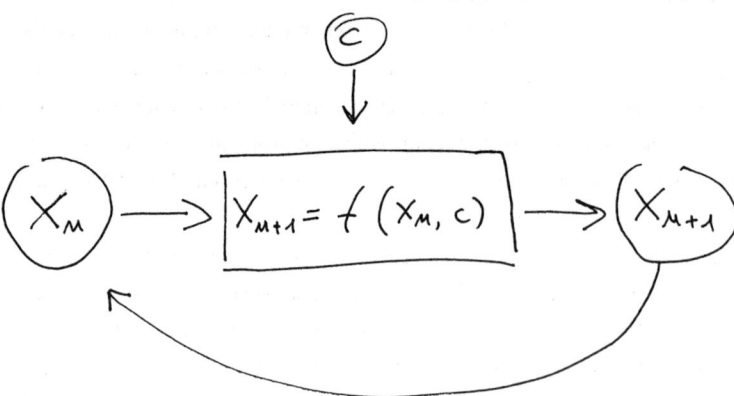

The recursion may be deterministic, stochastic, or some combination of both. The
$(n+1)$th value of the recursive feature is a function of the nth value, so that even when
this is a linear function the relation over two or more generations must be non linear

(10).

Looking at the recursion depicted in Fig.3 we can easily appreciate the fact that almost any behavior of complex systems can be formulated in such a way, the only need being some relevant discretization of time course and the sensible dependance of the studied system from its history (i.e the presence of a dynamics).

This means that almost everything can be described as a fractal and thence that calling something a fractal is a statement almost completely devoid of any information content. Obviously, this is true if we limit ourselves to the simple recognition of a structure like the one of Fig.3, the things change if we give a characterization to f and c parameters of Fig.3 so becoming able to infer some otherwise hidden features of interest of the studied system.

3. Conclusion

The general feeling I tried to express with the above considerations is the need to consider non-linear dynamics derived concepts as useful data analysis tools allowing us to discover new and unexpected features of complex phenomena. This power can be fully exploited only if we resist to the temptation of too early and ambitious theoretical efforts and restrict ourselves to a metaphorical data analysis perspective with an attitude more similar to multivariate statistical analysis than to theoretical physics (7,11,13,36).

References

1.D. Amit. *Modelling Brain Function: The World of Attractor Neural Networks*. Cambridge: Cambridge Univ. Press.(1989)

2. J.A. Anderson *et al. Psych Rev.* **5 (84)**, 413 (1977).

3. H.Atlan *Entre le Cristal et la Fumee*. Paris: Editions du Seuil (1979).

4. H.Atlan H. and I.R. Cohen (1989) *Theories of Immune Networks*. Springer-Verlag Heidelberg.

5. A.Babloyantz and A. Destexhe *Proc.Natl.Acad.Sci. USA* **(83)**, 3513 (1986).

6. P.Baldi and K. Hornik *Neural Networks*, **(2)**,53 (1989) .

7. S.Banerjee *et al. Journal of Theoretical Biology*, **(143)**, 91 (1990).

8. D.J. Bartholomew *Biometrika*, **(71)**, 221 (1984)

9. E. Basar *Am J. Physiol.*, **(245)**, R510 (1983)

10. J.B. Bassingthwaighte *Circulation Research* **(65)**: 578 (1989)

11. R. Benigni and A. Giuliani. *Am. Journ. of Physiol.* **(266) (35)** R1697 (1994)

12. J.P. Benzecri, *La pratique del l' analyse des donnes*. Paris: Dunod. (1980)

13. D.S. Broomhead and G.P. King. *Physica* **(20D)** 217 (1986)

14. G. Careri *Ordine e disordine nella materia*. Atti dell' Accademia Nazionale dei Lincei (1981)

15. S.Carra' *La Formazione delle Strutture*. Bollati Boringhieri, Torino.(1989)

16. D.R. Chialvo, *et al. Nature*, **(343)** 653 (1990)

17. J.J Collins and C.J. DeLuca *Phys. Rev. Lett.* **(73) (5)** 764 (1994)

18. M. Curé and J.P. Rolinat *Physiol. and Behav.* **(51)**, 771 (1992)

19. A. Delcour *et al J. Neurosci.* **(13)**: 181 (1993)

20. W. Ebenhoh *Theor. Pop. Biol.* **(42) (2)**, 152 (1992).

21. J.P Eckmann *et al* *Europhys. Lett.* **4 (9)** 973 (1987).

22. D.S. Faber *et al.. Science*, **(258)** 1494 (1992).

23. H. Frauenfelder *Nature structural biology* **(2) (10)**: 821 (1995)

24. H. Frauenfelder *et al Science*, **(254)**: 1598 (1991).

25. H. Frauenfelder and P.G. Wolynes *Physics Today* **(2),** 58 (1994)

26. M. Ghil, and R. Vautard *Nature*, **(350)**: 324 (1991)

27. A. Giuliani *et al. Neurobiol. of Learning and Memory* **(65)**: 82 (1996).

28. A. Giuliani and C. Manetti *Phys. Rev. E*, **(53) (6)**: 6336 (1996)

29. A. Giuliani *et al. Biol. Cybern.* **(74)**: 181 (1996)

30. J.J. Hopfield *Proc. Natl. Acad. Sci. USA*, **(79),** 2554 (1982).

31. N. Hubalik *et al.* in *Stress: Neuroendocrine and Molecular Approaches*, 977 (1992)

32. G. Iori *et al. Europhys. Lett.* **(25) (7):** 491 (1994)

33. S.A. Kauffmann, *Sci.Am.*, **(8),** 64 (1991)

34. P.R. Killeen *Behav. and Brain Sciences* **(17) (1)** 105 (1994)

35. S. Kirkpatrick and B. Selman *Science* **(264)** 1297 (1994)

36. L. Lebart *et al. Multivariate descriptive statistical analysis.* New York: Wiley. (1984)

37. D. Lerner *Qualita' e Quantita' ed altre Categorie della Scienza.* Boringhieri, Torino. (1971)

38. S.A. Levin *et al. Science* **(275)**: 334 (1997).

39. G. Nicolis and I. Prigogine *Exploring Complexity. An Introduction.* Munchen: R. Piper Gmbh KG. (1987)

40. R.W. Preisendorfer *Principal Component Analysis in Meteorology and Oceanography. Development in Atmospheric Science*, **17**, Amsterdam: Elsevier. (1988)

41. D.A. Rand and H.B. Wilson (1995). *Proc. R. Soc. Lond. B* **(259):** 111 (1995)

42. D.L. Stein *Sci. Am.* (**7**): 36 (1989).

43. L.L. Trulla *et al.* *Physics Letters A* (**223**): 255 (1996) .

44. J.M . Van Rossum and J.E.G. De Bie *TiPS,* (**12**), 379 (1991)

45. R. Vautard *et al.* *Physica* (**58D**), 95 (1992)

46. C.L. Webber and J.P. Zbilut *J. Appl.Physiol.* **76(2)**: 965 (1994)

47.   C.L. Webber. and J.P. Zbilut in: *Bioengineering Approaches to Pulmonary Physiology and Medicine* Khoo Plenum Press, New York: pp. 137 (1996)

48. K. Wiesenfeld K. and F. Moss *Nature* (**373**): 33 (1995)

49. K.G. Wilson *Sci.Am.* (**8**), 140 (1979)

50. P.G. Wolynes *et al.. Science* (**267**): 1619 .(1995)

51. M. Zak *Complex Syst.* (**7**): 59 (1993)